WINNING THE LONG COMPETITION

ALAN PENTZ

WINNING

THE

LONG

COMPETITION

THE KEY TO THE
NEXT AMERICAN CENTURY

LIONCREST
PUBLISHING

WINNING THE LONG COMPETITION

The Key to the Next American Century

ISBN   978-1-5445-1677-6   *Paperback*

978-1-5445-1676-9   *Ebook*

978-1-5445-1678-3   *Audiobook*

CONTENTS

INTRODUCTION

IT WAS MARCH 13, 2004. A CROWD OF SCIENTISTS, ENGI-neers, and students stood in the Mojave Desert outside Barstow, California, flanked by a fleet of futuristic-looking vehicles. Tension filled the air as the participants scrambled to prepare their vehicles—checking and re-checking connections, making last-minute adjustments, and scanning the long, winding course stretched out before them.

The Grand Challenge, the first-ever official race between autonomous vehicles, was about to begin.

This landmark moment had been three years in the making. In 2001, wanting to keep soldiers away from harm and combat zones, the United States Congress introduced an initiative to make a full third of all military ground combat vehicles uncrewed by 2015. It was a promising initiative, but one that quickly fell behind Congress's ambitious goals.

This wasn't for lack of trying. The usual defense industry stalwarts—General Dynamics, Northrup, Lockheed Martin, and Raytheon—put their full weight behind the project. But despite their efforts, they couldn't innovate quickly enough to develop the sensor and computing technologies necessary to make autonomous vehicles work.

To make up for lost ground, in February 2003, Tony Tether, Director of the Defense Advanced Research Projects Agency (DARPA), announced the Grand Challenge. This 142-mile race—open to anyone—offered a $1 million prize to the team whose self-driving car could finish the course the fastest. After a qualifying round of twenty-five participants at the California Speedway in Fontana, fifteen teams were selected for the Grand Challenge, most from high-profile universities like Carnegie Mellon and Stanford.

Finally, on that fateful day in the Mojave Desert, it was time to test their efforts. The only question remaining was who would walk away with the million-dollar prize.

As it turned out, no one.

Just three hours into a planned ten-hour competition, every single vehicle had failed somewhere along the track. Some rolled over onto their sides and were left to languish in the desert heat. Others detected nonexistent obstacles and swerved wildly off course to avoid them. Still others had unexpected breakdowns that left them dead on the track. When the dust had settled and all was said and done, only four vehicles remained operational.

This wasn't how any of the participants had expected the race to go. But such is the nature of innovation.

Undeterred, DARPA hosted a follow-up Grand Challenge in 2005, upping the bounty to $2 million as further incentive. Its perseverance paid off. This time, five teams finished the 132-mile course, with Stanford's team taking the top prize. The 2007 participants in the Grand Challenge fared even better, with six teams finishing the race.

Despite its rocky start, the era of autonomous vehicles was now officially underway.

More innovation followed. By 2009, seeing the potential in an expansive new industry, the private sector threw its hat into the ring. First Google announced its own autonomous vehicle program, and then companies like General Motors, Ford, and Uber joined in. To lead their fledgling programs, these companies hired the best and brightest in the business—who just so happened to be the very same engineers who had put their blood, sweat, and tears into winning the Grand Challenge just a few years prior. Over the next decade, the autonomous vehicle industry has grown rapidly—so rapidly, in fact, that it is projected to grow into a $7 trillion-dollar industry by 2050.[1] Not bad for an initial investment of $2 million![2]

THE AMERICAN FORMULA FOR SUCCESS

| THE AMERICAN FORMULA | NEW OPPORTUNITY ARISES | GOVERNMENT INVESTS RESOURCES | PRIVATE SECTOR INNOVATES |

The Grand Challenge represents America's potent formula for success—a formula that has driven innovation and economic success since the founding of the nation.

The basic formula is simple:

1. A new technological or economic opportunity arises.
2. The government invests in the basic research and infrastructure to help build an ecosystem around that opportunity.

---

1 Roger Lancot. *Accelerating the Future: The Economic Impact of the Emerging Passenger Economy.* Strategy Analytics (June 2017), https://newsroom.intel.com/newsroom/wp-content/uploads/sites/11/2017/05/passenger-economy.pdf.

2 For more on the history of the Grand Challenge, see: Alex Davis. "An Oral History of the DARPA Grand Challenge, the Grueling Robot Race That Launched the Self-Driving Car." *Wired.* August 3, 2017, https://www.wired.com/story/darpa-grand-challenge-2004-oral-history/.

3. The private sector then steps in to innovate and develop that ecosystem.

As the story of the Grand Challenge demonstrates, this formula has many advantages.

ADVANTAGE #1: IT ATTRACTS THE BEST AND THE BRIGHTEST

Government investment can convene people, companies, nonprofits, and academia, and drive innovation in a way that is more difficult or cost prohibitive for commercial interests. A prestigious government prize has a way of convening the best and the brightest when a near-term economic interest doesn't exist. Plenty of people had been working on different aspects of autonomous technology prior to the Grand Challenge, but it took the government's challenge to bring them together to compete and learn from each other. That provided the critical spark of innovation for autonomy.

This process often involves bringing new people to the game. One notable element of the Grand Challenge was the absence of the largest, most well-known defense contractors of the time. Instead, the effort was driven by the university system and Silicon Valley.

New innovations often welcome new players to the game. The old-school contractors took a shot at autonomous driving, but they focused too much on upgrading current technology instead of embracing new approaches. For the autonomous vehicle industry to move forward, new perspectives were required.

ADVANTAGE #2: IT JUMPSTARTS NEW INDUSTRIES

The government is ideally suited to take on such high-risk ventures—especially when it stands to benefit from the innovations those ventures produce. In the case of autonomous vehicles, Congress had a clear need—specifically, uncrewed combat vehicles. By footing the bill for the initial investment, the government was also able to kick-start the commercial autonomous vehicle industry.

The government does this in two ways. First, it acts as an early customer, allowing risky and unproven technologies access to a steady flow of money while those companies develop commercial businesses. Second, it can create open systems, standards, and specifications on which others can innovate. Just look at how the internet has grown and evolved over the past several decades. The government provided the initial investment and then stepped aside to let private industry innovate. The same process played out with autonomous vehicles: the government provided the incentive to compete, and some seed capital that then attracted investment from Google, Tesla, Uber, Ford, GM, and many others.

ADVANTAGE #3: IT GENERATES TREMENDOUS RETURN ON INVESTMENT (ROI)

As the Grand Challenge shows, a small amount of money can lead to big change. With a $2 million investment in 2005, the government gave life to what is projected to become a multitrillion-dollar industry.

Although this investment might have been relatively small, it was crucial to drive innovation in an emerging field. Autonomous technology was risky and unproven at the time of the Grand Challenge, with commercial applications still years away. Without that initial investment—and the results it produced—private investors would have shied away.

WE'VE REACHED A TURNING POINT

For centuries, the American Formula has spurred technological and economic growth. It gave us the railroad system, the highway system, the space program, the computer, and the internet, just to name a few examples. Each has represented a major turning point in our history, further establishing America as the primary economic leader on the global stage.

Today, we're at another major technological turning point as the cloud, artificial intelligence, and autonomy create a world in which innovations like autonomous vehicles will cross over from the fringe to the mainstream. New industries will be born, new entrepreneurial geniuses will emerge, and our way of life will change in new and exciting ways.

However, despite the promise of these developments, America's role within it is in question. Today, two major challenges confront us, and our response will largely determine our place in the new digital economy.

CHALLENGE #1: AMERICAN INVESTMENT IS DECLINING

The American Formula may have propelled us to the forefront of the global stage, but ever since the end of the Cold War in 1991, the United States has slowly turned away from this formula, deemphasizing investment in research and development (R&D) and infrastructure.

Why? One reason is that since the fall of the Soviet Union, America has lacked a true competitor. Without the need to measure ourselves against another nation, investing in new, high-risk innovations lost its appeal. As a result, government-funded R&D has become an increasingly smaller percentage of gross domestic product falling from a high of 2.3 percent of GDP in the 1960s to a low of just over 0.6 percent in 2020.[3]

It's true that while government investment shrank, private

---

3   Caleb Foote. "Federal Support for R&D Continues Its Ignominious Slide." Information
     Technology & Innovation Foundation. August 12, 2019, https://itif.org/publications/2019/08/12/
     federal-support-rd-continues-its-ignominious-slide.

sector investment grew. In many ways, this is a good thing. Why should taxpayers have to foot the bill? The problem is the private sector tends to invest in lower-risk areas with high profit potential. In essence, they are commercializing existing research. More fundamental research is expensive and difficult to link to commercial applications for sometimes decades. Government research also tends to favor open technology and systems that allow many innovators to leverage any investment. Beyond research, basic infrastructure is a public good. No one company wants to invest in something everyone can use.

This is not the position we want to be in as technology develops. If America is to maintain its economic and political position in the twenty-first century, it must invest heavily in new technologies like 5G, AI, and biogenetics as the crucial building blocks of the twenty-first century economy. While American companies remain innovative, we can only expect so much from private investors and shareholders. Relying solely on the private sector to fund innovation will also ensure that only the largest and most dominant companies succeed. By investing in and helping to build new platforms, the government can also help encourage new competition— and therefore, more innovation.

Despite turning away from its proven formula for success, America can still succeed in the twenty-first century economy. In the twentieth century, we built the businesses, we had the money, we had the leading technology, we had the power, and we had the influence. As a result, America got to enjoy the largest economy and the biggest stock market in the world.

But as we move from 4G to 5G, a lack of investment can change that situation. Whoever builds the best infrastructure the fastest will win. With an emerging new competitor to the east, America may find its success is not guaranteed.

CHALLENGE #2: CHINESE INNOVATION IS RISING

In 2015, Chinese Premier, Li Keqiang, released a plan called Made in China 2025. This ambitious plan laid out a clear road map for achieving technological dominance over the course of the next decade. The initiative marks the latest efforts by the Chinese government to grow their technological and economic capacity and cement their position on the global stage.

For years, America and American companies were happy to have China as a trade partner. Seeing an opportunity for previously untapped markets, companies like Cisco happily stepped in to help China with its technological efforts, such as building the Great Firewall.[4]

More recently, this friendly cooperation has begun to sour:

---

4   Sarah Lai Stirland. "Cisco Leak: 'Great Firewall' of China Was a Chance to Sell More Routers." *Wired.* May 20, 2008, https://www.wired.com/2008/05/leaked-cisco-do/.

1. The Chinese government is making big investments in building their own 5G network and subsidizing companies like Huawei to build Western networks. If Huawei becomes the only competitive infrastructure builder, this poses a significant security concern for Western countries and companies.[5]

2. American technology is being stolen in an effort to supersede American technological supremacy.[6]

3. American companies like Google and Facebook are all but shut out in China, even while Chinese companies like TikTok operate without a hitch in American markets.

4. Chinese companies like Alibaba and Tencent have become powerful rivals to their Western counterparts.

5. Chinese supply lines are seen as increasingly unreliable due to pandemics like COVID-19 and trade battles with the United States and other partners.

On the one hand, you can't blame China for their efforts. China is merely a hungry, upstart nation acting the way most countries would in a similar situation. Its people were poor, are increasingly middle income, and want to be rich. It has suffered for a century unable to project power, and its government wants more. (In fact, as we'll see in Part 2, America

---

5   Ryan Mcmorrow. "Huawei a Key Beneficiary of China Subsidies that US Wants Ended. PhysOrg. May 30, 2019, https://phys.org/news/2019-05-huawei-key-beneficiary-china-subsidies.html.

6   Sherisse Pham. "How Much Has the US Lost from China's IP Theft?" CNN Business. March 23, 2018, https://money.cnn.com/2018/03/23/technology/china-us-trump-tariffs-ip-theft/index.html.

behaved almost the exact same way toward Great Britain in the 1800s.)

Whatever the case, the result is clear: China is making early and massive investments to get its 5G network up and running and advance in critical future technologies, and America could potentially fall behind. If we don't get back to doing what we do best, we will lose the technological and economic advantage we have enjoyed since the early twentieth century.

WELCOME TO THE LONG COMPETITION

America may be falling behind, but we're not out of this race quite yet. In fact, our Long Competition with China has just begun.

At the outset of the Cold War in 1947, American diplomat, George Kennan, realized that the Soviet economic and political systems were flawed and unsustainable. All America had to do to win the Cold War was contain the threat and let the Soviet Union collapse under its own weight.

This strategy won't work in the Long Competition. The Long Competition is neither a hot nor a cold war. Though America and China's relationship may be contentious at times, we still enjoy a robust political and trade relationship. Instead of a binary war over which economic system will dominate, the Long Competition is about a continually shifting landscape of technological supremacy, which leads to economic supremacy, which leads to military supremacy. The key to winning that competition is embracing the American Formula and getting back to what America does best.

SOME CAVEATS

This book is a policy framework for winning the Long Competition. At the core of this framework is a return to the American Formula—specifically, an increased government investment in core current technology areas that require large-scale infrastructure build-outs and into high-risk R&D with underdeveloped or uncertain business models. By investing in core technology infrastructure like 5G and in the unproven, high-risk technologies that will take advantage of that infrastructure, the government creates an open, stable foundation on which the private sector can build.

This is a bipartisan framework for success. As such, it is neither pro-big government nor pro-big business. While I am advocating for the government to have a role in this process, I am not advocating for new bureaucracies. We have the tools in place to manage this effort. Further, I am also not advocating for giveaways to the private sector to make big business bigger. In fact, I believe if this effort is managed correctly, it can help diversify the ranks of successful business. I am also not taking any stance on an antitrust policy or the idea of breaking up large businesses. Those policy issues are beyond the scope of my argument.

China represents America's chief rival in the Long Competition, but in no way do I intend this book as anti-Chinese. Nor is it my intention to stir alarmism or to demonize the Chinese government or the Chinese people. As an American, I've written a book that offers a framework for America to maintain its own technological and economic standing in the world. However, I fully recognize why the Chinese government would want the same for its people. Nothing in this book is meant to imply there is anything inherently wrong with China pursuing its own interests. My argument is simply that they are a competitor and should be recognized as such.

MY BACKGROUND

This book draws on my more than two decades of experience working first on Capitol Hill, with the federal bureaucracy, and in the private sector as the founder and owner of a company. In my early career in public service, first as a legislative assistant for a congressman, and then as a speechwriter for US Senator (and later ambassador to China) Max Baucus, I learned the ins and outs of policymaking.

Of course, policymaking is one thing. Executing it is another. For over a decade, my consultancy, Corner Alliance, has worked with government leaders in the R&D and innovation communities across the Departments of Homeland Security, Defense, Commerce, National Institutes of Health, state and local government, and the nonprofit sector. This work has taught me how to put these ideas into practice in a way that is structured, measurable, and achievable—and that positions America for sustained success in the Long Competition. This book also represents my evolution in thinking about China. I, like many others, believed that engagement and trade with China would be a win-win for both countries and for the world. China would grow into a responsible power and gradually reform its political system, and the United States would benefit from a new global partner and open up a vast new market for its economy.

Those beliefs were all clearly wrong and were a misreading of history. First, history teaches us that when a new power rises, there is an inevitable tension and conflict with the existing powers. As a history major, I should have known better than to buy the Pollyanna views of the 1990s and early 2000s foreign policy and economic elites. The faster China rose, the more inevitable its competition with the United States became. Second, China's political system is fragile and historically has suffered from periods of breakdown and regional conflict. China's Communist Party perceives political reform as a gateway to state failure and their own loss of power similar to the collapse of the Soviet Union.[7] Third, our trade and the accom-

7   Wang Xiangwei. "Xi Jinping's Speech Shows China's Communist Party Is Still Haunted by the Fall of the Soviet Union." *SCMP*. April 6, 2019, https://www.scmp.com/week-asia/opinion/article/3004897/xi-jinpings-speech-shows-chinas-communist-party-still-haunted.

panying capital flows from China have had unequal effects on the US's own economy causing domestic inequality to rise.

These realizations have led me to believe that the US needs a new strategy to deal with a rising China and its own internal issues. That strategy must focus on doing what the US does best: innovating. My contention is that US innovation has always been a partnership between the government and the private sector, and that we need to return to that formula to win the competition with China and address our issues at home.

To lay out our framework for winning the Long Competition, I've organized this book into three parts:

1. Part 1 explores where we are today. In order to plot our course forward, we must first understand our starting point and what that means in terms of our advantages and disadvantages.
2. Part 2 dives into America's rich history of government/ private sector collaboration. While this is by no means a history book, it's important that we glance back to a few key moments in our history to understand what makes the American Formula so effective.
3. Part 3 charts our path forward. Once we understand both where we are and where we've been, we'll use those lessons to establish our priorities, identify key challenges, and lay out an agenda for moving forward.

Throughout my career, I've had the opportunity to witness, understand, and participate in the lead-up to this particular moment in history from a variety of perspectives. However, if there has been one common thread, it's this: The American Formula, when properly applied, is a powerful force for innovation and prosperity.

We don't need radical change to win the Long Competition. We already have the tools, the know-how, and the resources at our disposal. We just need to remember what we've done well and double-down on that process.

If we can do that, we can win the Long Competition.

PART 1

UNDERSTANDING THE CURRENT MOMENT

CHAPTER 1

WHERE ARE WE TODAY?

BEFORE THE CLOUD TOOK OFF, BEFORE MOBILITY GOT smart, there was Apple co-founder Steve Jobs.

While often a polarizing figure, Jobs was unquestionably good at stoking anticipation for new Apple products. So when the company announced that Jobs would be unveiling its newest product since the iPod at Apple's Worldwide Developer's Conference on June 29, 2007, rumors began to fly about Apple's first phone.

Expectations for the event were almost comically high. Many began referring to Apple's unknown device as "The Jesus Phone." No one knew exactly what to expect. Would Apple's new phone be a hit, or would it be dead on arrival?

When June 29th finally came, Jobs took the stage in the packed theater, and the crowd fell silent.

After briefly summarizing Apple' s history of innovation, Jobs got to the point. "Today, we're introducing three revolutionary products of this class [of innovative products]. The first one is a widescreen iPod with touch controls. The second is a revolutionary mobile phone. And the third is a breakthrough internet communications device."

The crowd began to stir.

"So, three things: a widescreen iPod with touch controls, a revolutionary mobile phone, and a breakthrough internet communications device."

Jobs paused again.

"An iPod, a phone, and an internet communicator."

The audience began breaking into applause.

"An iPod, a phone…are you getting it? These are not three separate devices! This is one device, and we are calling it iPhone! Today, Apple is going to reinvent the phone, and here it is."

By this point, Jobs had the entire room eating out of the

palm of his hand. Even today, watching the video almost feels like a religious experience.[8] In that moment, the era of the smartphone—and soon the cloud—had officially begun.

In many ways, the iPhone tells the story of the 2010s, a time when previously separate technologies began to blend together to spawn the mobile, cloud-based economy we enjoy today. Today, the iPhone is significant not just for its technological convergence, but also for how it perfectly encapsulates the American Formula.

Four essential components of the iPhone all began as projects for the US government. The Global Positioning System (GPS), the internet, Siri, and the multi-touchscreen were all funded by the Department of Defense and the National Science Foundation (NSF). That doesn't even account for the fact that the microchips it runs on were largely funded in the initial stages by the Department of Defense or that Apple's eventual biggest rival, Google, was originally funded by another NSF grant.

These technologies, first developed and championed by the United States government, eventually became available to private enterprise. When they did, they set the stage for private enterprise visionaries (like Jobs or Google's Sergey Brin and Larry Page) to create revolutionary new products and

---

8   Jonathan Turetta. "Steve Jobs iPhone 2007 Presentation (HD)." YouTube. May 13, 2013, https://www.youtube.com/watch?v=vN4U5FqrOdQ.

services that would propel the world of mobile and cloud computing. Companies like Apple, Facebook, Google, and Microsoft all led the way, consolidating technologies and developing exciting new devices and platforms at a rapid pace.

This era of rapid innovation wouldn't have been possible without the huge boost in capabilities made possible by 3G and then newly introduced 4G networks. While it was plain for anyone to see the enormous potential of this early smartphone, it was 4G that drove these devices and made further innovations possible. As we move into the 5G era, we're on the precipice of another technological turning point. Although we don't yet know what specific innovations it will inspire, the broad picture is already coming into focus. In this chapter, we explore the implications of a 5G world—both its potential and its challenges—from an American perspective.

UNDERSTANDING 5G

Throughout this book, we're going to be spending a lot of time discussing 5G networks and the many considerations necessary to make them a reality. So, before we go any further, it's important to define this term and explain how we'll be using it in the context of this book.

In many ways, words like 3G, 4G, and 5G aren't much more than marketing terms. No two 5G networks are the same—

though they do share some basic traits and technologies. Network carriers like AT&T and Verizon in America, as well as their counterparts from the industrialized world and their equipment manufacturers (Ericsson, Huawei, etc.), came together through the Third Generation Partnership Project (3GPP) to create a set of parameters that everyone agrees upon. (Confusingly, "3GPP" refers to the standards body that helped to develop technologies like 3G, 4G or LTE, and now 5G.) Over many years, committees of engineers work together to define what the next generation of wireless network systems will look like and how to build them.

The term 5G, then, simply means "fifth generation." With 5G, service providers all agree to work within a certain set of specifications—such as which chips to use in a phone or device, and which spectrum bands to use—but from there, they're free to design and deploy their technologies however they see fit. As a result, one 5G network implementation can and will be different from another.

Over the course of a given generation—be it 3G, 4G, or 5G, that carrier will update their network several times. In other words, there won't be a day when a switch will get flipped, and we will magically have 5G. It will be a process that will develop over several years.

Despite these differences in approach and implementation, we can still sketch out the basics of how 5G works and how it will likely be built out. 5G builds on top of all the 4G infrastructure already in place: cell towers, base stations, backhaul, etc. However, unlike 4G, which uses mid-spectrum bandwidths to send and receive data, 5G also uses higher bandwidths (30 GHz and above). These high bandwidths are well-suited for carrying high volumes of data and heavy traffic, but their range is limited. That limited range requires a lot of micro antennae or small cells to bridge the signal back to the larger towers that currently power our cellular communication. Building out America's 5G infrastructure, then, will require us to install a robust network of small devices—block by block and streetlamp by streetlamp—across the nation to seamlessly relay our data and keep all our devices talking to each other.

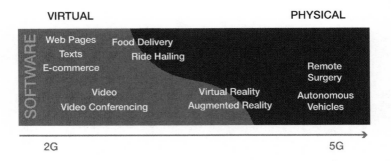

WHY 5G IS IMPORTANT

The key driver of the modern economy is data and the ability to compute with that data. As the world goes increasingly digital and becomes ever more dependent on the cloud, the ability to compute data quickly and at massive scale is the name of the game. When 4G was introduced, it came with throughput speeds ten times faster than 3G. The leap from 4G to 5G will be even greater, with estimated average speeds ranging from ten to a hundred times faster.[9]

With data moving at a hundred times the speed of 4G, the process of disruption begun in the 2010s will accelerate. Just like we saw after the iPhone was introduced, this rapid acceleration will give rise to new businesses and products that were previously unimaginable just two, five, or ten years before.

For this reason, we can't entirely predict the 5G future we'll soon be stepping into, though it will almost certainly involve tremendous growth in the following areas:

1. Autonomous vehicles
2. The Internet of Things (IoT)
3. Home automation
4. Wearables
5. Smart cities and infrastructure
6. Artificial Intelligence
7. Virtual Reality and Augmented Reality
8. Advanced biogenetics
9. Robotics

---

9   Angelo Ilumba. "5G vs. 4G vs. 3G: Comparing Generations of Mobile Network Technology." *WhistleOut.* March 25, 2019, https://www.whistleout.com/CellPhones/Guides/5g-vs-4g-vs-3g.

5G will also allow companies to complete the disruptions of the 4G era.

DID SOFTWARE EAT THE WORLD?

In 2011, Marc Andreessen wrote a seminal op-ed in *The Wall Street Journal* called "Why Software is Eating the World."[10] The piece embodied the subsequent era of cloud-based companies taking over the economy and leading a rebound in the US stock markets after the Great Financial Crisis. Software enabled by rent-as-needed servers from Amazon Web Services; widespread deployment and adoption of broadband, 3G, and then 4G; and the rise of the smartphone dramatically increased the reach and profitability of software companies. One need only look at the top ten companies as measured by market capitalization in 2008 and 2020.

---

10   Marc Andreessen. "Why Software Is Eating the World." *The Wall Street Journal.* August 20, 2011, https://www.wsj.com/articles/SB10001424053111903480904576512250915629460.

RANK	2008		2020	
1	Exxon Mobil	406,067	Microsoft	1,200,000
2	PetroChina	259,836	Apple Inc.	1,113,000
3	Wal-Mart	219,898	Amazon.com	970,590
4	China Mobile	201,291	Alphabet Inc.	799,180
5	Procter & Gamble	184,576	Alibaba Group	521,740
6	ICBC	173,930	Facebook, Inc.	475,460
7	Microsoft	172,929	Tencent	471,660
8	AT&T	167,950	Berkshire Hathaway	440,830
9	Johnson & Johnson	166,002	Visa	357,020
10	General Electric	161,278	Johnson & Johnson	345,700

In 2008, the most valuable companies were in energy and financial services. The only software company on the list was Microsoft, which was milking a legacy Windows monopoly. Microsoft had yet to implement its very successful and profitable cloud strategy. By the time Andreessen wrote his op-ed in 2011, Apple had climbed almost to the top of the charts, while Microsoft held on by its fingernails (again, that cloud strategy had yet to be implemented).

By March 31, 2020, as the COVID-19 pandemic took hold of the world, the top seven most valuable companies were all primarily software companies (one could argue about Apple), and two of them were also Chinese, demonstrating the country's rapid rise (see chapter 2). We've also seen many other startups emerge as unicorns rocketing to billions of dollars in revenue as incumbent (i.e., non-software-based) players see their businesses wither.

Do these developments mean that Andreessen was right? Did software really eat the world? Well, not quite all of it.

In April 2020, Andreessen released a blog called "It's Time to Build."[11] In it, he makes the case that the COVID-19 pandemic had revealed the failure of software to eat all of the world. While the price of running a web application had fallen 99 percent over the last decade, the cost of many other goods and services had done nothing but grow. Healthcare, housing, and education have continued to increase far beyond the rate of inflation, and for many, these goods are now nearly unaffordable. Additionally, our cities, transportation systems, and other infrastructures all suffer from a lack of investment.

Why is this? While the virtual world of the internet, mobility, and the cloud has thrived even as it has ridden down the cost curve, more traditional, physical, protected, and regulated industries have taken more of a share of our collective wallets (healthcare, education) or suffered from underinvestment (infrastructure).

---

11   Marc Andreessen. "It's Time to Build." Andreessen Horowitz. April 18, 2020, https://a16z.
       com/2020/04/18/its-time-to-build/.

Andreessen's answer for why this happened is a lack of will. A lack of desire to overcome "not in my backyard" development or rent-seeking higher educational institutions. A lack of desire to invest in twenty-first century infrastructure. I agree, but only in part. What I see is primarily another chapter in an ongoing story of technological investment and progress. Software hasn't eaten the whole world, but it also isn't full yet.

TECHNOLOGICAL PROGRESS IS NONLINEAR

In the early 2000s, the world saw the internet tech bubble grow and then pop. Companies like Pets.com rose quickly and then disappeared even faster, but while their innovative business models wowed the investing world and inflated their shares, most were never able to deliver on their initial promise.

For companies like Pets.com, their products and services looked great on paper. However, the infrastructure necessary to support their core business model wasn't yet in place. They may have had innovative ideas and were able to see where the world was headed, but they were too early in internet adoption to translate their ideas into long-term success. But while a company like Pets.com couldn't gain traction in 2000, by 2020 with ubiquitous smartphones, higher adoption of e-commerce, and more robust logistics networks, their core model had been adopted to great success by companies like Chewy. As Andreessen pointed out, many of the business models in the dot-com bust were later revived to great success (see Figure 1.1).[12]

FIGURE 1.1: FAILED DOTCOM COMPANIES WHOSE MODELS WERE RESURRECTED IN THE 2010S

Early 2000s	Modern Equivalent
Pets.com	Chewy, Wag
Webvan.com, Kozmo, Urbanfetch	Walmart, Whole Foods, Peapod, Shipt, Fresh Direct
Flooz, Beenz, SpeedyBucks	Bitcoin, Libra, Tether, Ethereum
Broadcast.com	TuneIn, Live365, Accuradio, iHeartRadio, podcasting
Digital Entertainment Network	Netflix, Hulu, Disney+, Apple TV+
Napster, Ritmoteca, Riffage	Spotify, Pandora, Apple Music and iTunes, YouTube Music
Beautyjungle.com	Hoseanna, GuyHaus, Birchbox
GovWorks	Services exist throughout the country
GeoCities	WordPress, Squarespace, WIX, Shopify

This is the boom, bust, and deployment model outlined by Carlota Perez in her book *Technological Revolutions and Financial Capital*. At first, new innovations are greeted with a boom, where companies and startups flood the market with business offerings related to the innovation. Often, however, it's too much, too soon; neither the infrastructure nor the market is developed enough to support so much competition. This inevitably leads to a bust where the majority of

---

12 Ryan Cohen. "Chewy Founder: We're no Pets.com." CNBC. July 26, 2019, https://www.cnbc.com/2019/07/26/opinion-chewy-is-no-petscom.html.

these businesses fail, but because the underlying technology or innovation is sound, their models are successfully resurrected years later when the core infrastructure has been deployed.

The second wave of 4G-enabled companies in the 2010s saw a similar grow-and-pop phenomenon in the areas where software met the physical world. Companies like WeWork were the highest-valued tech startup one moment, and close to filing for bankruptcy the next.[13] Other tech-enabled companies like the food delivery service Blue Apron have seen similar struggles.[14]

The boom we saw in startups that brought more technology and software to the physical world in the 2010s were the harbingers of the future—just as Pets.com was the harbinger of Chewy.com, and the 4G-enabled platforms like Netflix, Spotify, Zoom, and Instagram. It also enabled Uber and Grubhub, though those companies have struggled to turn a profit. 5G has the potential to solve many of these problems.[15] If we succeed in developing autonomous vehicles, Uber is suddenly profitable, because it doesn't have to

---

13   Dakin Campbell. "How WeWork Spiraled from a $47 Billion Valuation to Talk of Bankruptcy in Just 6
       Weeks." *Business Insider*. September 28, 2019, https://www.businessinsider.com/weworks-nightmare-ipo.

14   Brittain Ladd. "Blue Apron: What's Killing the Iconic Meal-Kit Company?"
       Forbes. August 2, 2018, https://www.forbes.com/sites/brittainladd/2018/08/02/
       killing-blue-apron-can-anything-save-the-iconic-meal-kit-company/#43488f19660d.

15   I am using 5G as a catch-all term for many technologies associated with the Fourth Industrial Revolution:
       AI, Virtual Reality, 5G, Machine Learning, etc.

pay drivers.[16] If we build out a network of ghost kitchens as well, Grubhub can cut out the restaurant and the driver.[17] This applies to your doctor as well. If a surgeon can operate remotely through robotic machines on a patient hundreds or thousands of miles away, that surgeon's productivity goes through the roof.[18] Additionally, if that doctor can replace 40 percent of her job using an AI bot to answer basic questions or perform a diagnosis, then that doctor can spend 40 percent of her time doing higher-value things. That's how you increase productivity.

This applies to academia as well. If a college professor can stream his or her course to thousands or even millions of students, his or her relative productivity has just skyrocketed. While a Cousera course today might seem like a pale reflection of your seminar class on nineteenth-century French poetry, it's only because you're thinking of the technology and tools driving those courses in a 4G context. 5G will create virtual experiences unlike anything we see today. While we certainly will not replace all in-person education

---

16 A quick note on jobs here. Many believe that technology will eliminate all the jobs, and we will have to put everyone on a universal basic income. No doubt technology eliminates jobs, but it creates others. When I graduated from college in the mid-1990s, you could get a job in data entry, but no one had ever heard of a social media strategist.

17 Roaming Hunger. "Guide to Ghost Kitchens (2020): All You Need to Know." February 3, 2020, https:// roaminghunger.com/blog/15623/ghost-kitchens-everything-you-must-know.

18 Caroline Frost. "5G Is Being Used to Perform Remote Surgery from Thousands of Miles Away, and It Could Transform the Healthcare Industry." *Business Insider.* August 16, 2019, https://www.businessinsider. com/5g-surgery-could-transform-healthcare-industry-2019-8.

on college campuses, we will open high quality education to many more students at a vastly reduced cost.

5G will allow software to eat much more of the physical world. Industries that have been safe until now—healthcare, education, construction, etc.—will be disrupted. That will be great for some, like the college professor who will make three times as much serving one hundred times as many students. It won't be so good for others, such as the also-ran college professor who cannot compete. Most importantly, it will be infinitely better for society as a whole, bringing these high-cost services to more people.

So, like Marc Andreessen, I agree that we need to build 5G and the technologies that it will enable. It's time to enter the next phase of technological development. It's also crucial that our government partner with the private sector by investing in that 5G infrastructure and R&D to ensure that the United States leads the world in these new industries. That kind of investment will deliver the results implied by companies like Uber, and it will deliver a new American century. If we don't make these investments, China will—and America's preeminent position will diminish.

CHAPTER 2

AN ASCENDANT CHINA?

TODAY, WE STAND AT THE DAWN OF THE 5G ERA. THESE are exciting times, but they are also uncertain ones. American companies and the American economy were the clear winners of the 4G revolution. This success is due in large part to the investments the US government made in the twentieth century—in technologies like microchips, wireless networks, the internet, and the global positioning system—which have benefitted us for many years.

However, while American companies have enjoyed unquestionable success during this period, Chinese companies have as well. Alibaba, Baidu, Huawei, and TenCent emerged on the global economic stage over the last decade. These companies' success has helped China emerge as a major economic and technological player in the 2010s.

As we turn our attention to the 5G era, America and China find themselves on increasingly equal footing. America's advantage is eroding, and we are facing a tough competitor who is determined to win. As American investment in R&D, infrastructure, and innovation has dwindled, China's is rising. It's as if the United States and China have been running an 800-meter race. While America jumped out ahead in the first lap, we slowed down as we came around for the second lap, just as China found its stride. China is now closing in from behind, and it is more than willing to invest the time, energy, and money to supplant us as the global economic and technological leader in the new 5G landscape. In this chapter, we will explore the mechanisms that led to an ascendant China, and what that might mean for America's position on the global stage in an uncertain future.

THE MADE IN CHINA 2025 INITIATIVE

In his book *Stealth War*, General Robert Spalding, former military attaché to the United States Embassy in Beijing, describes how the Chinese have been focused on gaining technological and military dominance for years, but they have done so under the radar.[19] They've kept it quiet. The architect of China's embrace of capitalism, Deng Xiaoping,

19   Robert Spalding. *Stealth War.* 2019. New York: Penguin Random House.

once said, "Hide your [China's] strength, bide your [China's] time."[20] It was a dictum that Xiaoping and his successors followed with great success, but China's current leader, Xi Jinping, seems to have abandoned it altogether.

Recently, Spaulding argues that America has finally begun to realize the implications of China's once quiet ascendency. US political leaders like Senator Marco Rubio have begun to warn about the dangers of China gaining ground in cutting-edge technologies.[21] We have also seen Senator Chuck Schumer advocating for new and massive federal investments in artificial intelligence to remain competitive with the Chinese.[22] Both Schumer and Senator Tom Cotton joined together to call for an investigation into TikTok and Chinese influence. Anytime you see Senators Schumer and Cotton both signing off on the same letter, you know the issue is truly bipartisan.[23]

---

20   Tobin Harshaw and Kevin Rudd. "Emperor Xi's China Is Done Biding Its Time." Harvard Kennedy School—Belfer Center. March 3, 2018, https://www.belfercenter.org/publication/emperor-xis-china-done-biding-its-time.

21   Marco Rubio. "Rubio Releases Report Outlining China's Plan for Global Dominance and Why America Must Respond." February 12, 2019, https://www.rubio.senate.gov/public/index.cfm/2019/2/rubio-releases-report-outlining-china-s-plan-for-global-dominance-and-why-america-must-respond.

22   Jeffrey Mervis. "United States Should Make a Massive Investment in AI, Top Senate Democrat Says." *Science*. November 11, 2019, https://www.sciencemag.org/news/2019/11/united-states-should-make-massive-investment-ai-top-senate-democrat-says.

23   Senate Democrats. "Bipartisan Lawmakers Urge President Trump to Oppose China's Bid to Lead the UN World Intellectual Property Org., Say China's Repeated Violations of Intellectual Property Protections Disqualifying." December 18, 2019, https://www.democrats.senate.gov/newsroom/press-releases/biparchinaschct.

In the twentieth century, China's gains had mostly been in labor-intensive industries and low-skill manufacturing. However, since the dawn of the twenty-first century, due to the increasing costs of labor, increasing wealth, and a workforce that is expected to shrink, China has focused on transitioning to a higher technology, services, and value-added economy.[24]

Given how transformational 5G will be, it is clear that whatever country and set of companies make the largest and smartest investments will have a significant economic and consequent military advantage. China appears highly aware of the opportunity posed by 5G and other advanced technologies; it is motivated to take advantage of it; and it has a strategy in place to do so. The "Made in China 2025" initiative, based on Germany's Industry 4.0 development plan, outlines China's dream for a new, tech-driven future.[25]

The China 2025 program lays out ten industries where it wants to achieve autonomy and make others dependent on it:

- Information technology (such as AI, smart devices, and Internet of Things [IoT])
- Robotics
- Machine learning

24 Euan McKirdy. "Study: China Faces 'Unstoppable' Population Decline by Mid-Century." CNN World. January 7, 2019, https://www.cnn.com/2019/01/07/asia/china-population-decline-study-intl/index.html.

25 China State Council. *Made in China 2025*. July 7, 2015, http://www.cittadellascienza.it/cina/wp-content/uploads/2017/02/IoT-ONE-Made-in-China-2025.pdf.

- Green energy
- Green vehicles (which are autonomous)
- Aerospace equipment
- Ocean engineering and high-tech ships
- Railway and power equipment
- Medicine and medical devices
- Agricultural machinery

While China 2025 is a written plan, it is also a general concept. As time has gone on, new technologies like 5G and AI have been added to the list. Dominating these industries will enable China to make the most money, build the most wealth, and obtain the most power.

For several decades, China's economy has depended heavily on its trade relationship with America and the rest of the world, which has increasingly become a point of contention between the two nations. Now an ascendant power, China no longer wishes to be economically or technologically dependent on the United States.

China's strategy focuses on: (1) shutting American companies out of key sectors; and (2) creating its own suppliers for critical technologies (in many cases, the basis of which are often stolen from American companies).[26] By subsidizing

---

26 Asa Fitch and Dan Strumpf. "Huawei Manages to Make Smartphones Without American
    Chips." *The Wall Street Journal.* December 1, 2019, https://www.wsj.com/articles/
    huawei-manages-to-make-smartphones-without-american-chips-11575196201.

these industries and making large investments to grow them, China hopes to free themselves from technological dependence on the United States.

China's Great Firewall and the birth of the *Splinternet* serve as a good example. China's firewall was designed to severely restrict internet access within the country so that it could control the information most of its citizens receive and maintain the Communist Party's grip on power. In doing so, China essentially created an alternative—or splinter—internet. To bolster this splinternet, China shut out American companies like Google, favoring its own companies like Baidu instead. This move will increasingly lead to competing internets: the internet of the West, the internet of China, and other "belt-and-road" countries that are squarely in China's sphere of influence.

WEAPONIZED INTERDEPENDENCE

The ascent of China has until now been a story of integration with the world economy, but the splinternet shows us the path forward. As we've seen in the COVID-19 pandemic, supply lines and dependencies matter quite a bit. China is aware that dependence on other countries for key supplies can be dangerous—and that other countries' dependence on China is a powerful weapon. This growing awareness reflects a concept that Henry Farrell and Abraham L. Newman refer to as *weaponized interdependence.*[27] This concept has two key premises: First, if a country depends on another country for key resources, then there is an opening to weaponize that dependence. Second, a country should seek to minimize its own dependence.

For example, in May 2020, the Trump Administration banned the sale of high-end microchips made with American equipment to China that are essential for smartphones and 5G base stations. That ban will slow China's ability to dominate the 5G market. The administration's effort was a clear attempt to weaponize China's dependence on companies using American equipment to manufacture chips.[28] China attempted something similar with a rare earth element export ban it levied against Japan in a dispute in the early 2010s.[29]

To further unpack the concept of weaponized interdependence, let's look at how these interdependencies arise in the supply chain, using the recent COVID-19 crisis as an example. Even if countries do not intend to weaponize their interdependence, it still happens in resource-scarce environ-

27   Henry Farrell and Abraham L. Newman. "Weaponized Interdependence: How Global Economic Networks Shape State Coercion." International Security 44, no. 1 (Summer 2019), https://doi.org/10.1162/isec_a_00351.

28   Reuters. "US Prepares Crackdown on Huawei's Global Chip Supply, Sources Say." CNBC. March 26, 2020, https://www.cnbc.com/2020/03/27/us-prepares-crackdown-on-huaweis-global-chip-supply-sources-say.html.

29   Keith Bradsher. "Amid Tension, China Blocks Vital Exports to Japan." The New York Times. September 22, 2010, https://www.nytimes.com/2010/09/23/business/global/23rare.html.

ments. In recent years, China has been the world's leading manufacturer not only of ventilators, but also many of the active ingredients in many common pharmaceutical drugs. Once China began to understand the scope of the problem posed by the Coronavirus, the government implemented export controls on both the ventilators and certain chemical compounds—which resulted in a partial global breakdown of the global medical supply chain.[30]

To be clear, in this case, China didn't make this choice to punish or gain advantage over any other country. That is, they didn't weaponize other countries' dependence on them. They simply needed these resources for themselves in the midst of a massive public health crisis. Germany made a similar difficult choice in withholding medical supplies from Italy, a fellow member of the European Union. Germany recognized Italy's need, and they recognized that withholding these supplies would hurt Italy's medical response, but they also recognized their own need and chose to prioritize that.[31]

Now both China and the US realize the power of making others dependent on you and the weakness of being depen-

30 Kate O'Keeffe, Liza Lin, and Eva Xiao. "China's Export Restrictions Strand Medical Goods US Needs to Fight Coronavirus, State Department Says." *The Wall Street Journal*, April 16, 2020, https://www.wsj.com/articles/chinas-export-restrictions-strand-medical-goods-u-s-needs-to-fight-coronavirus-state-department-says-11587031203.

31 Francesco Guarascio and Philip Blenkinsop. "EU Fails to Persuade France, Germany to Life Coronavirus Health Gear Controls." *Reuters*. March 6, 2020, https://www.reuters.com/article/us-health-coronavirus-eu/eu-fails-to-persuade-france-germany-to-lift-coronavirus-health-gear-controls-idUSKBN20T166.

dent on others. It also largely mirrors the strategies of companies like Amazon, Google, and Alibaba. This realization is a large part of what will drive the intensity of the Long Competition. The story going forward will not be one of increasing integration, but of trade blocs, division, and competition.

A GROWING TENSION

The China 2025 initiative, America's own evolving economic interests, and a state of weaponized interdependence have given rise to a growing tension between the two countries.

As we have seen during the coronavirus crisis, China and the United States were quickly at each other's throats, flinging allegations back and forth. But the ill will preceded COVID-19. One particular headline-catching example involved a skirmish between the NBA and the Chinese government. In October 2019, Houston Rockets General Manager, Daryl Morey, posted a tweet in support of Chinese protesters in Hong Kong. Outraged by this tweet, the Chinese government began censoring NBA broadcasts, and then several Chinese companies suspended their deals with the NBA as well.[32]

This put the NBA in an unenviable position: defend a team owner's right to free speech and risk losing access to the

---

32   Mac Schneider. "China's Fight with the NBA, Explained." *Vox.* November 11, 2019, https://www.vox.com/videos/2019/11/11/20959250/china-nba-houston-rockets-hong-kong.

500 million Chinese viewers who regularly tuned into the NBA, or sanction/fire him and risk the condemnation of their home country for kowtowing to the new enemy? The NBA eventually sided with Morey, which cost the league an estimated $400 million.[33]

As the controversy surrounding this incident demonstrated, this wasn't an easy choice for anyone to make. However, other American companies—who once enjoyed considerable latitude in courting Chinese markets—have also found themselves in similar situations as the NBA. As these issues come to light in the public sphere, many Americans are increasingly unhappy with their companies' apparent willingness to censor themselves for the Chinese government.

Americans once believed a strategic partnership with China would emerge as the Chinese became wealthier. We believed if we invited the Chinese onto the world stage, they would become more like us, but the opposite has happened. As Chinese economic influence grows, China is increasingly leveraging that power to bend American companies to their will.

For years, American companies have largely tried to ignore the problem, hoping that it would all blow over soon enough.

33   Luis Paez-Pumar. "Commissioner Adam Silver Says NBA's Clash with China Cost 'Hundreds of Millions' of Dollars." *Inside Hook*. February 16, 2020, https://www.insidehook.com/daily_brief/sports/nba-china-adam-silver-daryl-morey.

Throw in an increasingly contentious trade war between the two countries and a pandemic that has raised serious questions about global supply chains, and it's time to realize this is a permanent issue.

THE ERA OF THE GLOBAL CITIZEN IS OVER

American corporations that conceived of themselves less as "American" and more as "global citizens" will need to readjust. Being a global citizen was easy during a time of American dominance. Corporations like Google and Amazon rose to prominence before China had emerged as a clear competitor. In such an environment, it was easy to expand to other markets.

But that period is over. China now poses a clear threat to our technological supremacy, and American companies must recognize that the balancing act between America and China is becoming increasingly difficult. In recent years, China has shut Google and Facebook out of their country in favor of their own companies. And it doesn't stop there.

As countless other companies have come to realize, from Starbucks to Apple, such is the new reality of playing "global citizen" between the US and China. You're damned if you do, and damned if you don't. The effects of this phenomena will only accelerate in the post-Coronavirus landscape, as critical supply chains in manufacturing and healthcare come under the scrutiny of governments worldwide. Areas we thought were open to free trade or just places we thought wouldn't be problematic are suddenly very problematic. The more its economy grows, the more China will insist on using its power to try to force Western companies to adopt its message, and the more pressure Western companies will come under from their governments and fellow citizens to resist that pressure.

Joseph Stalin once said, "If I could control the American motion picture, I would need nothing else to convert the entire world to communism." China has read its share of history, and it knows exactly what Stalin meant by this. However, they've also learned that they don't have to take Hollywood by force. They just have to threaten Hollywood's global distribution rights, and suddenly there are no Chinese villains in any movies—and Taiwan doesn't exist.[34]

Again, I don't blame the companies themselves. They want to make money for their shareholders, and China has plenty of money. While they are not in the business of global politics, bipartisan pressure from their home government and increasing demands from the Chinese government will leave many of these companies with no choice but to choose sides.

---

34 Inkoo Kang. "Hollywood's Changing Its Movies to Appease the Chinese? Good." *The Atlantic*. March 30, 2013, https://www.theatlantic.com/entertainment/archive/2013/03/hollywoods-changing-its-movies-to-appease-the-chinese-good/274174/.

TECHNOLOGY CONVERGENCE AND NATIONAL SECURITY

As the Daryl Morley case demonstrates, nationalist pressure will increasingly spread from traditional national security sectors to larger parts of the economy. The concept of technology convergence has exacerbated the tension between China and the United States. 4G- and 5G-based technologies have reduced everything into data and software, which has in turn blurred the line between what is critical national security technology and what isn't. Seemingly separate technology areas will continue to merge as they are fully digitized and automated. Apple is a good example of this convergence. It started as a computer company that revolutionized the mobile phone market, and then it decimated the camera industry, reshaped the music industry, and made Swiss watchmakers and headphone manufacturers a bunch of also-rans. The iPhone and related products merged traditionally separate verticals into one.

In this new playing field, the role of government entities like the Committee on Foreign Investment in the United States (CFIUS) has dramatically expanded. CFIUS is an inter-agency government committee run by the Department of the Treasury that reviews and regulates foreign investments in American businesses and assets—especially those with government and national security implications. For instance, CFIUS might review a French company's acquisition of a small defense contractor that is developing technology valuable to the American government. CFIUS might approve the purchase, but on the condition that the company creates a special board in the United States that is populated only by American citizens.

However, large sections of the American economy now find themselves involved in the Long Competition as a result of technology convergence. In this way, competition has moved well beyond what we would have traditionally called military applications or equipment.

As one example in recent years, the American government has banned 5G equipment manufactured by the Chinese company, Huawei—and they have called on the rest of the world to follow suit. This decision came after accusations that Huawei has or can use their equipment to spy on behalf of the Chinese government.[35]

This is new territory. We don't really have an analog in history with this level of complexity, and our current situation is pushing institutions like CFIUS to the limits of its existing mandate. For example, CFIUS responded to the Schumer/ Cotton letter and recently opened an investigation into the acquisition of the American company Musical.ly by the Chi-

---

35   Emily Feng. "China's Tech Giant Huawei Spans Much of the Globe Despite US Efforts to Ban It." NPR. October 24, 2019, https://www.npr.org/2019/10/24/759902041/ chinas-tech-giant-huawei-spans-much-of-the-globe-despite-u-s-efforts-to-ban-it.

nese company Byte Dance.[36] Musical.ly was a social media platform for creating lip syncing videos that is now part of the Byte Dance platform, TikTok. Examples like this demonstrate how we've left the realm of bombs and planes and entered a world where we're worried about foreign powers brainwashing our children with fifteen-second videos. Joe Stalin would be proud.

So even if America and China do work out a long-term trade deal, American companies would be wise to expect that another trade war or something similar might be just around the corner. It is a Long Competition, after all, and as such, the skirmishes could last for decades.

REVIEWING WHERE WE ARE

This is the situation we find ourselves in at the dawn of the 5G era. We're standing on the verge of an exciting new future, but at this moment that future is marked with uncertainty. America may have won the 4G era, but today we face a competitor who is hungry, coordinated, and focused.

In the future, American companies won't have the same access to the Chinese market and will face constant threats of confiscation of assets, arrest, and more subtle forms of

---

36 Greg Roumeliotis, Yingzhi Yang, Echo Wang, Alexandra Alper. "Exclusive: US Opens National Security Investigation into TikTok—Sources" Reuters. November 1, 2019, https://www.reuters.com/article/us-tiktok-cfius-exclusive/exclusive-u-s-opens-national-security-investigation-into-tiktok-sources-idUSKBN1XB4IL.

economic warfare if they don't comply with China's strict guidelines.[37] Additionally, Western companies will face increasing pressure from their governments not to sell critical technology to the Chinese.

In the face of such competition, the United States has to decide how it's going to play. To win, we must invest. The question is, can we make it happen? Is it too late to change course and remember our strengths? I don't think so. After all, especially when it comes to technology, a lot can change in just a short amount of time.

A few years ago, about ten years after the first iPhone came out, my family and I were driving to my in-laws' house in Hilton Head, South Carolina.

Somewhere in North Carolina, we stopped at a rest area. There, by the restrooms and the vending machines, we saw a pay phone.

"Do you know what this is?" I asked my kids.

They sort of knew, but they had never interacted with one— as their questions made obvious.

One of them asked where the money went, and then he asked

---

37   Daniel Shane. "How China Gets What It Wants from American Companies. CNN Business. April 5, 2018, https://money.cnn.com/2018/04/05/news/economy/china-foreign-companies-restrictions/index.html.

why the phone didn't accept credit cards. Another was confused why there was a cord attached to the headset. All three were fascinated by it.

Ten years is a relatively short period of time, not even a full generation. And yet, in such a short time, our nation had undergone such a massive change in technology that my kids barely even knew what a pay phone was.

5G will alter the playing field in much the same way. Right now, there's still time to take the reins and drive that change. However, in order to understand how to move forward, we must first look to the past to better understand our unique strength: the government private sector partnership that drives the American Formula.

In part 2, we will explore the evolution of this model, one in which the government invests in core areas of need to build out the necessary infrastructure and platforms that will allow innovators and entrepreneurs to thrive. As we explore this partnership, we will begin to understand what differentiates the United States from China—chiefly, how we are able to fund research and drive innovation without state control—and how that ultimately gives us an advantage in the Long Competition.

PART 2

LEARNING
FROM THE PAST

CHAPTER 3

THE FIRST INDUSTRIAL REVOLUTION: 1780–1870

IN PART 1, WE LAID OUT THE CURRENT STATE OF AMER-
ica's technological and economic position relative to China
and stakes involved in taking leadership in 5G. In short,
we're losing ground in the Long Competition, but we have a
unique advantage: the American Formula for success.

Throughout its history, the United States has used the
America Formula to create unparalleled technological and
economic prosperity. However, this formula wasn't designed
and perfected overnight. Through a series of iterations over

the nineteenth, twentieth, and twenty-first centuries, America has refined and honed its unique approach to innovation.

Over the last 250 years, the United States has gone from a small group of colonies hugging the eastern seaboard to a global superpower. During that time, the world itself has gone through multiple rounds of innovation. We commonly refer to the first as the Industrial Revolution and estimate that it began sometime in eighteenth-century Britain, but many scholars now think we have experienced not just one, but multiple industrial revolutions (IRs) from the eighteenth to the twenty-first centuries.

EXPANSION OF TECH

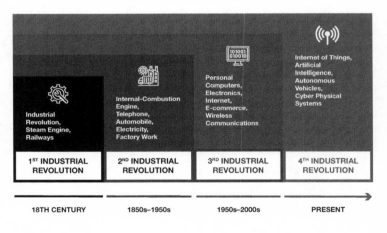

1ST INDUSTRIAL REVOLUTION	2ND INDUSTRIAL REVOLUTION	3RD INDUSTRIAL REVOLUTION	4TH INDUSTRIAL REVOLUTION
Industrial Revolution, Steam Engine, Railways	Internal-Combustion Engine, Telephone, Automobile, Electricity, Factory Work	Personal Computers, Electronics, Internet, E-commerce, Wireless Communications	Internet of Things, Artificial Intelligence, Autonomous Vehicles, Cyber Physical Systems
18TH CENTURY	1850s–1950s	1950s–2000s	PRESENT

The First IR was powered first by water and then by steam and early mechanization. During this era from the 1700s through the early 1800s, America was the follower, stealing technology from Great Britain, much as China has stolen from us in recent years. The Second IR was powered by electricity and saw the rise of mass production. During that period, America caught up and then passed the rest of the world. The Third IR marked the rise of computing, information technology, and automation. In the Third IR, the United States beat back the Soviet threat and established itself as the sole superpower. Today we stand on the precipice of the Fourth IR, in which virtual and physical systems will come together. Think of biogenetics, autonomous vehicles, AI, and the Internet of Things (IoT).

At this moment, the question is whether the United States will maintain its advantage, or whether China will displace it as the preeminent technological power. We're going to chart our journey to this moment weaving our way through the four industrial revolutions and contrasting them with the current environment so we can begin to map out a path forward.

THE EARLY DAYS

The America of the First IR was still very much in its infancy. American industry in particular was young, arrogant, and, shall we say, rough around the edges. Many American businessmen weren't afraid of playing dirty to get their way—often at the expense of established British and European interests. For example:

1. In 1787 an American agent named Andrew Mitchell, with close connections to the future US Secretary of the Treasury, Alexander Hamilton, had the plans for advanced British weaving machines seized from his luggage as he boarded a ship to depart England. He eventually escaped and fled to Denmark, but his plans were captured.

2. In a more successful attempt, an experienced British loom operator, named Samuel Slater, escaped from the country in search of the bounties he had heard were on offer in the United States. He teamed up with Moses Brown (uncle of Brown University's namesake) to create one of the most advanced textile operations in the world. It was considered intellectual property theft by the British, but Slater retired one of the richest men in America and has been credited as "the father of the American Industrial Revolution."[38]

3. Francis Cabot Lowell talked his way into British mills and managed to memorize much of the setup and machinery design. He returned to America and founded one of the first company towns dedicated to textile manufacturing in Lowell, Massachusetts.[39]

With the distance of history, these stories are equal parts amusing and surprising. Were American businesses really so flagrantly willing to flout British law and steal their intellec-

---

38 Wikipedia. "Samuel Slater." Visited: March 26, 2020, https://en.wikipedia.org/wiki/Samuel_Slater.

39 Ian Harvey. "The Sneaky Ways America First Lured Innovative "Thinkers and Industries to Its Shores." *The Vintage News*. September 2, 2018, https://www.thevintagenews.com/2018/09/02/industrial-espionage/.

tual property (IP)? Of course they were—just as China has been equally willing to steal American IP today. Just as the British lamented America's penchant for intellectual property theft, America would lament China's similar tendencies to create a black market for bootlegged music, movies, and video games—and later, more serious technology thefts that China would use to accelerate their growth at the beginning of the Fourth IR. If you don't think Americans stole, just look to Charles Dickens, who estimated he would double his income if the United States cracked down on pirated versions of his books.[40]

While America of the past and China of the present both embraced their kleptomaniacal sides, they were also busy establishing the infrastructure and economic models that would eventually earn them a more legitimate place at the global economic table.

In eighteenth-century America, this duty was taken up by Alexander Hamilton, who almost single-handedly laid the groundwork for the modern US economy. Without the robust manufacturing capacity that we enjoy today, the America of the First IR largely propped up its economy by selling commodities like tobacco and then cotton to Britain in return for Britain's more valuable manufactured goods.[41]

---

40   BBC News. "When Charles Dickens Fell Out with America." February 4, 2012, https://www.bbc.com/news/magazine-17017791.

41   Of course, this also involved the slave trade.

Because of this unbalanced exchange, most of America's cash was going back to Britain, leaving very little liquidity on this side of the Atlantic and most of our Founding Fathers were deeply in debt to British factors and merchants.

Seeing this economic disparity, Hamilton realized the newly independent America had some crucial needs:

1. It needed to show it could pay its debts so it could borrow additional money to invest.
2. It needed to establish financial markets and financial institutions.
3. It needed to invest in industry to move up the value chain from agricultural and commodity goods.

To address these needs, Hamilton released a series of reports. The first two, released in 1790 are the most famous. They dealt with public credit and the establishment of a national bank. They met resistance from James Madison and Thomas Jefferson and their block of Southern votes. After a deal in which Hamilton agreed to locate the new capital in today's Washington, DC, those proposals passed. The third, the Report on Manufactures, was released in 1791, was never fully adopted, though all of its most significant recommendations were put in place piecemeal. Specifically, the report called for "bounties" (aka, subsidies or prizes) to businesses to develop manufactured goods and tariffs to protect those nascent companies from foreign competition. From the moment these frameworks were adopted, the government became an active partner of the private sector, establishing a clearly defined role in forming financial markets and in spurring innovation and industry.[42] Historically, the government has done this in four ways:

1. **As a customer.** The government in the form of the US military and other agencies is a big-spending early customer for many technologies.

2. **As a funder of innovation.** The government provides grants or contracts to academia and industry to do basic research and development.

3. **As a funder of infrastructure.** The private sector may have built out our railroads, highways, and telecommunications capacities, but they did so with active funding and other incentives from the government.

4. **As a standards developer.** The government also works to create a set of standards and/or open systems for many parts of the economy. The founding of the internet serves as a good example. Before the internet as we know it, companies like the Digital Equipment Corporation (DEC) had their own proprietary technologies that connected their own equipment in networks. They were great innovations, but their proprietary nature limited their growth; developers would have had to pay the DEC

---

42 The New York Stock Exchange began as an agreement between brokers in "Hamilton Bonds" and other securities in 1792 who met under a buttonwood tree on Wall Street: https://time.com/4777959/buttonwood-agreement-stock-exchange/.

a fee to build on top of it. By establishing an architecture for an open internet, the government helped create a standard that everyone would benefit from.

Such an approach requires cooperation from both the government and the private sector to work. If the government developed new technologies or infrastructure on its own, those projects would never receive further investment from the private sector—nor would the government benefit from the ingenuity found in the private sector.

An early proof of concept of the American Formula at work can be seen in the story of the Erie Canal, a manmade passage that transformed commerce in the United States. Prior to the canal's construction, the only effective way to transport goods from east to west in the North was over the Appalachian Mountains. This was a dangerous and expensive process, slowing westward expansion and economic growth in the region to a crawl.

The situation only changed when a flour merchant went bankrupt after losing his shipment in the Appalachians and found himself in debtor's prison. Understandably upset at his misfortune, the merchant wrote angry letters to the editor of his local newspaper, complaining that he was a victim of circumstance and that merchants such as himself needed a viable way to transport goods west.

As the story goes, the governor of New York read the merchant's letters and, so moved, became a champion of the canal's construction. Eventually the governor had his way, and, by 1825, the Erie Canal was complete—allowing both goods and people to flow from the emerging Midwest to Albany and then down the Hudson River, the shipping center of New York City. The canal cemented New York's role as the main port, commercial center, and financial center for the United States, and set off a canal building boom that ended only with the emergence of the railroad. Not bad for a penniless flour merchant. In this instance, the investment necessary to take on such a massive engineering project was well outside the reach of private industry alone. By embracing its role as an engine of growth, the government created the conditions for industry to flourish.[43]

A NATION ON THE VERGE OF GREATNESS

The success of the Erie Canal would serve as a compelling proof of concept for the American Formula and the innovation to come. By the end of the First IR, America had successfully transitioned from a tech-stealing upstart to a nation ready to grow its industrial and commercial capacity and become an important player on the global stage.

As the era of the canal gave way to the era of the railroad, the United States began to take an increasingly competitive position that shifted to dominate by the First World War.

43   In this case, the state government.

Again, this transition was driven by the American Formula that took a largely agrarian nation divided by slavery and connected by unreliable roads through a bloody Civil War to victory in Europe and world power status in the wake of that war in just over fifty years. It was a truly astounding accomplishment. In the wake of the Civil War, the American Formula went into hyperdrive, knitting the country together in a network of railroads and building global businesses like US Steel, Standard Oil, General Electric, International Harvester, and JP Morgan & Co.

CHAPTER 4

THE SECOND INDUSTRIAL REVOLUTION (1870–1960)

IN THE FIRST INDUSTRIAL REVOLUTION, AMERICA MADE considerable progress for a fledgling nation. However, compared to Great Britain, the global economic leader at the time, it was still a small fish in a very large pond.

Driven by steam power, looms, and other society-changing inventions, Britain was riding high on a wave of manufacturing and exports. The United States had tobacco, cotton, farm products, lumber, and skins and pelts, but these products were all basic commodities. America still sat toward the lower end of the value chain.

In his *Report on Manufactures* (see previous chapter), Hamilton had foreseen this challenge, urging the nation to move beyond commodity exporting and develop a true domestic manufacturing capability. The government agreed, and throughout the early to mid-1800s, America began to shift from an agricultural to an industrial economy. In this chapter, we'll explore how this helped our young nation to win the Second Industrial Revolution.

TRAINS, TELEGRAPHS, AND AUTOMOBILES

During the Second Industrial Revolution, British investors were so flush with cash that they began pouring their excess capital into American industries and railroads. There were only so many railroads you could build in Britain, and the new world offered higher rates of return and massive opportunity. This onrush of capital paved the way for American growth and expansion throughout the nineteenth century.

Throughout the early 1800s, there was sectional tension over what was then called "internal improvements," or what we now call infrastructure. In the wake of the War of 1812, Henry Clay developed a set of policies he named "the American System." This three-point plan called for tariffs on imported goods from Europe, a national bank to stabilize financial markets, and funding for internal improvements. Southern planters were largely opposed to this plan. Their reason for doing so was simple: tariffs benefited domestic manufacturers who were largely located in the middle and northern part of the country, and they sold most of their crops to Europe. They knew the North and middle of the country would disproportionately benefit from the internal improvements and that retaliation from Europe would fall on America's main exports—their crops.

Later, during the Civil War with the Southern states in secession, President Abraham Lincoln seized his opportunity to break the stalemate. With the Pacific Railroad Act of 1862, Lincoln focused American investment on internal improvements, including the railroad system. Through this act, the government provided cash and land grants for private companies—chiefly Union Pacific and Central Pacific—to finally build out the transcontinental railroad.

This newly invigorated railroad system knit together a new continental nation and set the stage for the United States to assume global leadership. With the North dominating in the wake of the Civil War, this trend continued. British capital flooded into the country to build railroads and the industries of the future. This was the Gilded Age and the time of the robber barons who built the US steel, oil, shipping, and finance industries.

Running parallel to the railroads (both literally and figuratively) was another new innovation: the telegraph. In the 1830s, inventor Samuel Morse began privately developing the single-wire telegraph system. Seeing the potential of this invention to revolutionize the way America communicated, the US government became an early patron of Morse's work. In 1838, Morse gave the first public demonstration of his machine before Congress. By 1841, Morse had installed a telegraph system in the Capitol. Just a year later, pleased with the results, Congress granted Morse $30,000 to test the feasibility of creating a large-scale telegraph system.

Over the next several decades, the telegraph system would prove massively scalable. Wherever the railroads went, the telegraph followed.[44] Of course, such a robust communications network required skilled operators to send, receive, and translate messages. One of those operators was none other than future titan of industry, Andrew Carnegie.

Carnegie was dirt poor when he came to America from Scotland to be a bobbin boy in a Pittsburgh cotton mill. Eventually, Carnegie moved from bobbin boy to telegraph operator, where his ability to decipher messages without paper caught the eye of an administrator for the Pennsylvania Railroad. Carnegie soon became the administrator's own personal telegraph operator—and then the superintendent of the entire Western division of the railroad.

As Carnegie rapidly rose up the ranks at the railroad, more

---

44 History, Art, & Archives. "'What Hath God Wrought': The House and the Telegraph." https://history.house.gov/Exhibitions-and-Publications/Electronic-Technology/Telegraph/.

opportunities followed. Once he became an investor—first in a company that manufactured railroad cars—and then in a series of other promising industries, such as railroads and then steel—Carnegie was on his way to becoming the legendary tycoon he is known as today.[45]

Carnegie's story intersects with the stories of both the American railroad system, the telegraph, and the development of large-scale industrial businesses like steel. It also perfectly encapsulates America's journey from follower to dominant global player. First, the government seeded the ground by funding the telegraph and providing the land and legal rights to entrepreneurs who built the railroads. In this way, not only did the government's patronage help make the telegraph ubiquitous with the American railroad system, but it also paved the way for an innovative businessman like Carnegie to find the foothold that would catapult him to success. Carnegie differentiated himself by being skilled in a new technology—an opportunity he likely wouldn't have had if not for the government's full-throated support of this innovation.

Carnegie clawed his way up in business, making shrewd deals in the railroad industry that made him a wealthy man. During a trip to Britain in the post–Civil War era, Carnegie became familiar with a new process to make high quality

---

45 EyeWitness to History.com. "Andrew Carnegie Becomes a Capitalist, 1856." 2007, http://www. eyewitnesstohistory.com/carnegie.htm.

steel cheaply, called "the Bessemer Process."[46] He visited several production facilities and gathered as much information as he could. In the mid-1870s, he opened the first American steel mill using the new process and went on to dominate the US steel industry.[47] Francis Cabot Lowell would have been either proud or jealous. By 1889, US steel production outpaced Britain's production.[48] The follower had now become the leader.

Carnegie's ability to seize on opportunity has been echoed throughout American history, perhaps most notably in recent history by Bill Gates and Steve Jobs and their early adoption of government-subsidized computer technology to co-found Microsoft and Apple, respectively.

FROM UNDERDOG TO TOP DOG

The Second Industrial Revolution was a prolonged period of massive change both domestically and globally, bookended by the railroad system on one end and the highway system on the other. About midway through this period, the economic and military superiority of America and Great Britain changed hands.

---

46 Bessemer developed the process for the British military to make more accurate cannons. This is another example of government as a first customer to spur innovation in a massively important industry.

47 Robert J. McNamara. "The Bessemer Steel Process." ThoughtCo. April 2, 2019, https://www.thoughtco.com/bessemer-steel-process-definition-1773300.

48 Ellen Terrell. "Andrew Carnegie—Man of Steel." Library of Congress. December 3, 2012, https://blogs.loc.gov/inside_adams/2012/12/andrew-carnegie-man-of-steel/.

The British entered the Second Industrial Revolution in a strong position. Flush with cash, they began pouring resources into US industry in general and the US railroad system in particular. While this benefitted the investors, it would eventually hasten Britain's fall from the top of the global pecking order. By the turn of the twentieth century, it was clear to most that American economic, if not military, influence had become a serious competitor to that of Britain.

Most of this was probably inevitable based on the populations, geography, and resources of each country. It's tempting to draw a parallel between America's rise in the nineteenth century and China's rise today. The question of whether China surpasses America, however, as we once surpassed Britain, remains unclear.

For one, the United States was larger both in terms of land and population, so passing Britain was probably inevitable. Further, while America was focused on building across a largely contiguous piece of territory (a continent), Britain was stretched thin—both geographically and financially— tending to its far-flung empire. America also had a more cohesive population. While slavery and then segregation and mass immigration were major issues, one of which caused a bloody civil war, Britain's problems with ruling over the vast Indian subcontinent, for example, proved even more difficult. Britain had effectively been punching above its weight for two hundred years thanks to its colonial system and early embrace of the Industrial Revolution, but that era was rapidly drawing to a close even before the Second World War. The United States is much better positioned in this regard.

That said, it's clear to anyone watching that China is gaining ground quickly. If there is any comfort to take, once again it's in our unique way of doing things. In the American Formula, the government doesn't try to pick the winners in terms of any new innovation. For example, as Warren Buffett famously pointed out, every investor knew that *someone* was going to make a lot of money with the automobile. However, knowing which company would win was anybody's guess. The same was true of the telegraph: it might have looked like Samuel Morse was the primary beneficiary of the government's investment in the telegraph, but they couldn't possibly have known that a dirt-poor immigrant like Andrew Carnegie would amass far greater riches as at least a partial result of the technology.

One flaw in a centrally planned system like China's is that they try to pick winners and losers. While not an entirely centrally planned economy, China does have a large overlay of state-owned enterprises, which receive a majority of the official bank money, as well as considerable state support. This creates a powerful cadre of potentially inefficient companies that are propped up based on government influence, and in extreme cases kept alive only to provide jobs.

To be fair, the Chinese system does have a significant private sector for entrepreneurs. However, the moment one of these private-sector enterprises begins to succeed, the state becomes more involved, and the threat of government intervention is ever present. This might benefit the state, but it also stifles innovation in the process. As we will see in the competition between Russia and the United States during the Third IR, this inability to balance collaboration between the state and private enterprise (or the lack thereof) poses a major roadblock for sustained innovation.

CHAPTER 5

THE THIRD
REVOLUTION
(1960–2010)

"LO."

That was the very first internet message.

It was meant to be "logged in," but the fledgling TCP/IP system crashed. (For those of us who remember the days of dial-up, we can relate.)

Today, the internet has grown from a government-funded project to the backbone of our economy. As much of the world was essentially locked down during the COVID-19 pandemic of 2020, the internet has been essential in keeping at least some parts of the economy up and running. Aside from keeping us all connected and enabling many of us to do our jobs, it has also given us a treasure trove of cat videos, memes, and animated gifs.

The development of the internet is just one landmark moment in the Third Industrial Revolution. Chronologically, there is a bit of overlap between the Second and Third IRs. For instance, while Eisenhower's championing of the national highway system in the 1950s was certainly a Second IR outcome, the seeds of the Third Industrial Revolution began to take root some years earlier with the introduction of the ENIAC computer.

Despite the overlap in years, the Third IR is distinct from the Second in that it is characterized by the rise of digital and information technology (IT). Here is where America and the world first entered the age of computers—and where the American Formula produced some of its greatest successes.

THE BIRTH OF SILICON VALLEY

When William Shockley founded one of the world's first transistor companies in 1956, he had no idea he'd just cemented the sleepy town of Mountain View, California, as the birthplace of Silicon Valley.

If his investors had had their way, his company, Shockley Semiconductor Laboratory, would have set up shop about four hundred miles south in Newport Beach, California. However, wanting to be closer to his ailing mother, Shockley chose the San Francisco Bay Area instead.

Buoyed by his recently awarded Nobel Prize, Shockley was not only able to attract a crack team of scientists to work for him, but he was also able to secure the US military as his first customer. In fact, his Nobel Prize and many of his connections came from work at Bell Labs, which had close connections to the government through it's highly regulated monopoly parent, AT&T. For a while, things went well—that is, until it became obvious that Shockley was a terrible manager. Employees were often appalled at Shockley's paranoid behavior, and when they found out he'd been secretly recording their conversations, many were fed up.[49]

So, when proto-venture capitalist, Arthur Rock (he actually coined the term *venture capital*), approached several of these employees with an opportunity to start their own semiconductor company, they took it. Thus, with a $1.5 million investment and led by Shockley's "Traitorous Eight," Fairchild Semiconductor was born.

Fairchild quickly began edging in on Shockley's business, selling their transistors to weapons projects crucial to the early Cold War arms race. Soon, their transistors could be

49 Wikipedia. "William Shockley." Accessed February 26, 2020, https://en.wikipedia.org/wiki/William_Shockley.

found in the IBM-built computer inside the B-70, as well as in Autonetic's guidance system for the Minuteman ballistic missile.

In just a few short years, what began as a few determined semiconductor companies became a thriving industry, and Silicon Valley blossomed into the epicenter of the computing world that it is today. Some of those Traitorous Eight, including Gordon Moore (of Moore's Law fame) and Robert Noyce, left Fairchild to found Intel. Once again, we have the American Formula to thank. The efforts of private enterprises led by Shockley and Rock, combined with a series of government grants and military contracts, helped pave the way for an explosion of innovations that created the single greatest wealth-generating machine in history: Silicon Valley.

So why did the United States place such a high priority on developing the semiconductors? Simply put, competition is a great motivator. The early days of the microchip were also the early days of the Cold War, a time when both the Soviet Union and the United States raced to build increasingly advanced weaponry—and therefore a military advantage against their opponent.

The United States sought to gain this advantage through computers. One of our first attempts, the ENIAC Computer (also the product of a government grant), was a start, but it was also unreliable. Housed in the University of Pennsylvania, this massive vacuum tube-powered computer had a tendency to overheat. That, combined with the constant need to replace the tubes, led to a system that only worked about half the time. The military needed computers that were smaller, cooler, and more reliable. Eventually, they turned their attention to Shockley and Fairchild, whose work on semiconductors would produce powerful microchips more suited to the military's purposes.

Certainly, the money the government pumped into developing this technology would have been worth it if only to win the Cold War. However, because of America's robust economy, entrepreneurial activity, and deep financial markets, it also led to an industry that has created trillions of dollars of wealth and has infiltrated almost every facet of our lives.

IBM released the first fully transistorized commercial calculator in 1955. Hewlett-Packard followed with theirs in the late fifties and early sixties, and with that, the age of the computer was well underway. The personal computer came in the late seventies and eighties, followed by the World Wide Web, which made the internet far more usable and widely available by the mid-nineties and launched the era of software dominance. Eventually, that innovation led to the cloud, which has been essential to modern commerce. Today, computing devices are everywhere, powered by the mobility and fast speeds of 4G—which has set the stage for the dawning Fourth Industrial Revolution.

MAINTAINING ADVANTAGE

During the Third IR, the government was both a primary customer and primary investor in early computer and microchip technology. They funded research at Bell Labs. They funded the first fully functioning computer at the University of Pennsylvania. They funded the contracts that helped grow companies like IBM and Hewlett-Packard.

This massive outpouring of investments throughout the fifties, sixties, and seventies rapidly accelerated the introduction of the personal computer and the internet. However, as is often the case, the government wasn't as interested in the consumer applications as they were in the defense and military applications.

During the Third IR, the United States entered the Cold War, shifting its primary competitor from World War II-era Germany and Japan to the Soviet Union. As has been well-documented, the Cold War was a new kind of conflict, one fought not on an open battlefield, but through proxy wars and, even more importantly, through an ongoing race to achieve technological and military advantage. America pursued this strategy through a series of what are referred to as offsets, which tend to make the cost of competition cheaper.

In all, there were three significant offsets that helped America maintain its advantage in the Third IR.

THE FIRST OFFSET

The first offset involved the development of nuclear weapons at the end of World War II. By sheer numbers—soldiers, weapons, tanks, etc.—the Soviets had both America and Europe at a disadvantage at the outset of the Cold War. This was important, since America didn't want to spend the resources to match the Soviets in a pure arms race, the nations of Europe were devastated and focused on rebuilding their respective economies, and the United States had long supply chains across the Atlantic Ocean. We could have lost a conventional fight with the Soviets in Europe in 1945. Further, America didn't want to devote such a large portion of its economy and lives to fighting its new enemy in a conventional war.

Having nuclear weapons as an offset helped take the burden off the economy, enabling the private sector to continue to grow and thrive without the government crowding it out. Eventually, the Soviets developed a nuclear capability of their own, and in so doing, achieved strategic parity with the United States. Still, nuclear deterrence enabled the United States to bide its time building both its and Europe's economies and letting the Soviet Union slowly implode.

THE SECOND OFFSET

By the 1960s, America found its next offset in the form of digital technology. Advanced targeting computers gave America the capability to deploy its arsenal with far greater precision than the Russians. By the time of the Gulf War in the early 1990s, America had developed advanced smart weapons and stealth technology, further cementing their advantage (and leading to the development of GPS as a consumer product that underpins all our mapping, location, and air travel applications).

By this time, the Soviet Union had collapsed, and throughout the 1990s, America lacked a clear competitor. China's own economy was certainly developing during this time, but it would still be a decade or more before they would approach anything resembling technological or economic parity with the United States.

THE THIRD OFFSET

As the Cold War ended, government investment in R&D began to decline. With Russia no longer a major threat, and with America's military engagements largely taking place in nonstrategic areas of the globe against non-state or much less technologically sophisticated opponents, there was no longer any need for an offset. The United States was the only game in town, so we got a bit fat and happy—and then started throwing our weight around carelessly. But now that confidence and that seemingly insurmountable lead appears to be waning.

China has emerged as a new competitor, and once again

the United States is called to project power across the globe against an opponent with a far greater population. Now the battle is for the third offset: robotics, autonomy, advanced manufacturing, miniaturization, big data/AI, etc. That list of technologies bears an uncanny resemblance to the technologies China has targeted in Made in China 2025. The resemblance is no coincidence.

USING OFFSETS TO UNDERSTAND THE LONG COMPETITION

This process of offsets provides a useful lens for viewing the current competition with China. Just as Russia attempted to create parity with America throughout the Cold War, China is doing the same today. This is not to say we've entered a new Cold War with China. Several key differences make the current political and economic landscape less a Cold War and more a Long Competition. For instance:

1. The United States and China trade far more with each other than the United States ever did with the Soviet Union.
2. China is a technological power in a way that the Soviets never were.
3. China has a thriving semi-capitalist economy.

China will still exist even if it loses ground in the Long Competition.

Additionally, China isn't ideologically driven, and our main issue isn't with their particular form of government. They may favor autocracy, but they're not interested in turning the people of other countries into communists the way the Soviets were. China just wants what's good for China. A democratic China will want largely the same things. This means America isn't motivated to fight proxy wars across the globe to keep Chinese ideology at bay—nor should we be. If China wants to invest capital in Africa, that's not America's concern.

So, what *is* America's concern? The answer is twofold:

1. That China doesn't gain technological supremacy over the United States.
2. That China doesn't come to dominate their neighbors and region in Asia.

Making these our strategic priorities will have a far-reaching impact on American policy. First, it might cause America to reconsider its presence and involvement in the Middle East—specifically in countries such as Iraq, Libya, and Syria. While we have an interest in ensuring that no power can control the Middle East, our interests don't go any further than that. As an increasingly prominent oil producer, the United States is not dependent on Middle East oil anymore, and we simply want to prevent any one power from dominating the region. Elsewhere, we've already seen attempts to end

American involvement in Afghanistan, and let us hope we are gone from our longest war as soon as possible.[50]

Second, the United States will most likely continue its current trend toward withdrawing from Europe. Russia, with a population of less than a third of the European Union and GDP one quarter that of just Germany, hardly seems like a hegemonic threat. Third, this strategy avoids the mistake of chasing the Chinese around the world trying to counter every move. Post COVID-19 this is an increasing concern of mine. If the Chinese have to remain wary in their home region, we can maintain access to Asia, and prevent China from expending too many resources pursuing mischief in other regions—namely ours. But if they gain access to a port in Africa, we don't need to expend the time and resources worrying about it. We need to focus on keeping Japan, South Korea, India, etc. strong enough to give the Chinese pause, and keep them pinned down in their own neighborhood.

While we focus on gaining economic and technological supremacy, whoever does will be able to make the rules in the Fourth Industrial Revolution and reap its substantial benefits.

---

50   Matthew Lee and Kathy Gannon. "US and Taliban Sign Deal Aimed at Ending War in Afghanistan." *The Associated Press.* February 29, 2020, https://apnews.com/491544713df4879f399d0ff5523d369e.

CHAPTER 6

THE FOURTH REVOLUTION: THE PRESENT

IN 1991, AMERICA BEGAN CONSTRUCTION ON A SUPER-collider outside of Waxahachie, Texas. Just like the CERN supercollider that was later built near Geneva, Switzerland, the purpose of this supercollider would be to confirm the existence of the long-hypothesized Higgs Boson particle.

Over the next two years, the United States would pour $2 billion into the project—digging tunnels, managing designs, and developing the necessary technology to build such a massive machine. But then Congress pulled the plug, saying that the project had become too expensive to continue.

Meanwhile, the European Union continued to fund their Large Hadron Collider project at CERN, which began operating in 2008. Then, on July 4, 2012, researchers announced the discovery of the Higgs Boson, the biggest event in physics in over a generation—and just to add insult to injury, they did it on America's birthday.

This once-in-a-generation discovery could have been America's. However, by choosing frugality over research, both the discovery and America's reputation as the standard-bearer of high-energy physics went to Europe. All America has to show for its $2 billion investment is the skeleton of a 20 percent complete tunnel in the Texas desert—which they had to fill with water in order to maintain its structural integrity.[51] Even worse, according to a 1994 survey, the abandoned supercollider project also led to an unfortunate brain drain; nearly half of the scientists involved had since left the field of physics.[52] The brain drain that led the world's leading scientist to flee Nazi Germany's aggression and help build the United States' lead reversed.

The supercollider project itself wasn't a part of the Fourth Industrial Revolution, but it is an example of the government's failure to consistently invest in research and

51   Trevor English. "The USA's Super Collider Lies Abandoned in the Texas Desert." *Interesting Engineering.* June 7, 2016, https://interestingengineering.com/usas-super-collider-lies-abandoned-texas-desert.

52   Eric Berger. "Super Collider Lab Now Gathering Weeds Near Dallas." *Houston Chronicle.* May 25, 2008, https://www.chron.com/news/houston-texas/article/Super-Collider-Lab-Now-gathering-weeds-near-Dallas-1757185.php.

innovation after the Cold War ended in the early 1990s. After the Soviet Union collapsed, America celebrated its victory and then spent trillions fighting wars against non-state actors and minor states in the Middle East and Afghanistan. Without a true competitor or ideological threat, the American government no longer felt innovation was worth the cost. Many felt the government was the problem, and that the private sector was best equipped to spur innovation. With the rise of the internet and cloud computing, who could argue? Never mind that much of that innovation in the private sector was the result of public investment in previous decades.

We won the Third IR in part because we were motivated by the threat of the Soviet Union. The fear of Sputnik in 1957 became a catalyst for America to ramp up its own space efforts, first with the Saturn program, and then the Apollo program. For years, there was this uneasy feeling of roughly technological equivalency. However, that feeling changed to a sense of technological dominance once the Apollo 11 astronauts landed on the moon in 1969. Over the next couple of decades, from 1970 to 1990, America gained tremendous ground, and by the time the Soviet Union collapsed, it was apparent that the US had considerable technological superiority.

America had been riding high on that good feeling up until recently, when we collectively looked around and realized that China was coming up right behind us—and fast. This is how the stage is set for the coming Fourth Industrial Revolution. While the American Formula helped us win the past two IRs, nothing is guaranteed for the future.

Fortunately, while our drive has diminished, America is still innovating. Here in the last chapter of part 2, we'll examine the areas in which the American Formula is still benefiting us in order to begin plotting our course for the future.

MAPPING THE HUMAN GENOME

Part of the Fourth IR will be the blending of the physical and digital worlds. One big arena for this is biology—specifically genetics. Here, the American Formula has remained a constant, if occasionally somewhat contentious, force.

In 1998, Craig Venter became the founder, president, and chief scientific officer of the newly founded CELERA Genomics. The goal of CELERA was to sequence the human genome— and in so doing, become the definitive source of genomic and related medical and biological information. This sparked a rivalry with the team running the Human Genome Project (HGP) at the National Institutes of Health (NIH), which was, not coincidentally, Venter's most recent employer. From the moment Venter founded CELERA, the race was on to be the first to sequence the full human genome.

Venter took on this competition intentionally, believing that the HGP was taking too long, was too costly, and was getting bogged down by a less efficient approach. His goal was to complete the task by 2001—and to make a tidy profit while he was at it, since the first person to map the genome would also own it. Venter planned not only to file the preliminary patents for over 6,000 genes, but also to make the full human genome available only to paying customers.

The government was horrified at the idea that Venter might end up taking proprietary ownership of so much valuable information. By 2000, the bitter competition between Venter and the Human Genome Project became so heated that the White House intervened, asking the two parties to settle their differences and publish their findings jointly. They complied, and later that year they published their draft of the full human genome.

Venter and the HGP competed, made each other improve, and got the job done quicker than either had initially anticipated. Venter may have played the role of the bad guy, but his involvement ultimately accelerated the science of biogenetics—a major cornerstone of the Fourth IR. He saw inefficiency in the HGP's approach, he decided he could do it better, and he did. In turn, the government saw what Venter was doing and adjusted their own approach. While acrimonious at times, this public-private relationship ultimately benefited all parties involved.[53]

---

53   YourGenome.org. "Why Was There a Race to Sequence the Human Genome?" June 13, 2016, https://www. yourgenome.org/stories/why-was-there-a-race-to-sequence-the-human-genome.

The Fourth IR will blend the world of data and processing power with the physical world of biology. Cloud computing allows the storage and a sharing of the massive amounts of data involved in genetics. Technologies like CRISPR allow scientists to edit that genome quickly and cheaply, and various AI are emerging that will help us find new solutions to medical and genetic issues that we never could have discovered with human cognition alone. For example, researchers have used AI engines to identify likely treatments for COVID-19.[54] By remaining active with their work in the NIH, the government helped fund and spur the research for a complete revolution in medicine. The leader of the HGP, Dr. Francis Collins, currently leads the NIH as of this writing.

Even since this breakthrough in 2000, the government has been an active investor in the field of genetic research. The trick now is to translate that enthusiasm to other areas also in desperate need of a boost.

INNOVATION NEEDS BOTH GOVERNMENT AND THE PRIVATE SECTOR

We began this book with the story of DARPA and the autonomous vehicle. The government's investment here is another strong example of the American Formula driving innovations that will be crucial in the Fourth IR. Again, what makes

---

54   Jonathan Block. "COVID-19 Puts Spotlight on Artificial Intelligence." *GenEdge*. May 11, 2020, https://www.genengnews.com/gen-edge/covid-19-puts-spotlight-on-artificial-intelligence/.

this particular story so remarkable is that with a relatively small initial investment (a $2 million prize), the government was able to kickstart the entire autonomous vehicle industry—which is projected to be a $7 trillion industry by 2050.

Unfortunately, while this is an uplifting story, it's also all too rare. As I mentioned in part 1, government investment in R&D has been in decline even as private investment is increasing. The CELERA example shows the dangers of this approach. CELERA almost ended up patenting the human genome, but once it proved difficult to profit from, CELERA turned to other more promising business opportunities rather than continuing with basic research.

The same company that had accelerated genetics research abandoned the basic science and focused on applied science instead—a logical move for the company and its investors, but potentially not an optimal one for the nation.

Again, there is nothing wrong with companies pursuing profit; commercially viable applications of a new innovation are vital to the success of the American Formula. However, when only the sure things or near sure things receive funding, then the speculative opportunities—often the areas where true innovation originates—are left wanting.

Perhaps the only companies willing to invest in speculative R&D have been a few tech giants like Google, and they're often lampooned by analysts for "throwing their money away." But if Google is able to make bolder bets than most, it's because they're also cartoonishly rich. They can afford to take on a level of risk that 99.9 percent of other companies can't, making them the exception that proves the rule. We've also seen recent evidence that Google is enforcing more financial discipline on those investments and closing down others. Eventually the market comes for the dreamers. The fact that Tesla still exists as the most shorted stock in history is again the exception that proves the rule. There aren't enough Elon Musks to drive speculative private sector innovation, and remember, there wouldn't be a Tesla without DARPA, or a SpaceX without NASA.

This isn't to imply that more companies should be like Google and take greater risks with speculative research. That's not a fair expectation. Companies should continue to do what companies do best. However, while the private sector is ill-equipped to take an active role in speculative research, the government is perfectly suited for that task—and it has a long, proud history of doing exactly that.

Because the government doesn't need to turn a profit, it is in a far better position than the private sector to play the long game. Just as in the past, these investments can help build the necessary infrastructure—technological or otherwise—that will benefit everyone down the road. As history shows, the profit will come eventually, but private enterprises and their investors often don't have the patience to wait.

While the twenty-first century has seen its share of exciting breakthroughs like the Human Genome Project, America could lose its leadership position as the driver of innovation to China. If this trend continues, stories like CERN, where we got cold feet and let someone else have the glory and benefits, will become more common.

The good news is that the American Formula remains unique on the global stage. If we can remember what made us successful and recommit to the approach that has served us so well, we can position ourselves for success in a way that China so far has been unable to replicate. It's a long road ahead, but in part 3, we'll begin to map it out.

PART 3

SETTING THE AGENDA

CHAPTER 7

EMBRACING AMERICA'S UNIQUE STRENGTHS

IN 1982, THE FRENCH GOVERNMENT, THROUGH FRANCE Telecom, introduced the Minitel. These boxy, beige devices were early networked computer systems, complete with keyboards, that connected to the phone line through a proprietary technology—and every French citizen got one.

The Minitel was incredibly advanced for its time. Through preloaded applications, users could check their stocks, conduct online banking, and perform other similar activities characteristic of the early internet. When Minitel first rolled out, it was incredibly successful, putting France ahead of the pack on the nascent information superhighway. Even as late as 1997, the Minitel was still the pride of France, with then-president Jacques Chirac famously saying, "Today a baker in Aubervilliers knows perfectly how to check his bank account on the Minitel. Can the same be said of the baker in New York?"[55]

At the time, Chirac had every right to boast. Although the internet is ubiquitous now, at the time, it still hadn't fully penetrated American markets. This was soon to change, however, and just a few years after Chirac had praised the Minitel, the French public had all but forgotten about it. By this time, the internet as we know it had taken off, and the Minitel had become woefully outdated. Still, the Minitel program stumbled on for several more years until it was finally put out of its misery in 2012.

France's grand experiment is interesting for a number of reasons. For starters, it gives us insight into the European model, which especially at the time was more state-driven than market-driven.[56] The French government hoped that by championing the Minitel and getting one into every household, their efforts would help spur innovation. But here, a funny thing happened. The newspapers all rebelled at the idea of the Minitel, certain that its mere existence was going to end their business. (They weren't entirely wrong, just a

---

55 Schofield, Hugh. 2012. "The Rise and Fall of The France-Wide Web." *BBC News*, https://www.bbc.com/news/magazine-18610692.

56 Ibid.

little premature. In the years that followed, the internet *would* dramatically reshape the newspaper business on a global scale.)

Fearful of their future, the newspapers began lobbying the government for protection. The government responded, granting the newspapers sole power to create apps for the Minitel. Such a move essentially made the government gate-keepers on behalf of the newspapers—and to the detriment of the Minitel. By barring all potential developers but newspapers from building out the Minitel's capabilities, they had effectively throttled innovation and sowed the seeds for the Minitel's inevitable demise.

When the government is allowed to control everything, it only incentivizes others to manipulate the government to their own ends—a fact that the French newspapers had now made painfully obvious, though their efforts ultimately backfired. Soon all kinds of businesses were registering as newspapers, just so they could secure the right to make applications. Even the porn companies joined in!

In the end, while the Minitel began life as an innovation ahead of its time and was considered a big win for the French government, today it serves as a cautionary tale of what can happen in a system that doesn't allow private enterprises to freely participate in new ventures in the spirit of continued innovation and creative destruction.

But could what happened with the Minitel happen here in America?

Not likely. As we'll see in this chapter, the American Formula has several advantages to the approach of other countries and governments. Several companies, like Digital Equipment Corporation (DEC), had proprietary networking systems. They failed and were left on the ash heap of history, because they were closed like Minitel. Our government funded open systems like TCP/IP and then allowed multiple players to experiment on top of those open standards. Eventually, despite its infamous attribution to Al Gore, the internet has been built by multiple players and remains resilient and adaptable as a result. The recent global lockdown from COVID-19 proved how resilient that system truly is. The American Formula may support long-term innovation, but it also accounts for the fact that innovation is brutal, and that a system unable to sustain itself long term must adapt or die. Just ask the investors in DEC.[57]

Here in part 3, we will return to the present state of innovation we explored in part 1 and begin drawing from the lessons we learned in part 2 to chart our path forward into the Fourth IR. In this chapter, we'll begin by exploring the

---

57   On a side note, DEC did adapt to the internet developing the early World Wide Web search engine, Alta Vista. While Alta Vista became one of the most visited sites on the nascent web, it eventually succumbed to creative destruction, as well as Google's search engine (originally funded by an NSF grant) cornered the search market in the 2000s. Innovation can be brutal.

nature of American innovation, and why historically it has been one of our greatest strengths.

THE AMERICAN ADVANTAGE

Now that we've had a chance to see the American Formula at work throughout our history, it's important that we understand what drives it. Here are the key elements that make the American Formula possible.

WORLD CLASS RESEARCH INSTITUTIONS

America's university and research system is the envy of the world. The United States has produced the most Nobel Prize winners by far and its universities dominate the world academic rankings. We funded land grant universities across the country in the nineteenth century and then seeded them with billions in R&D funding in the twentieth. It works. Government is no exception either. Scientists at NIST, Department of Energy Labs, and, of course, the NIH are on the cutting edge of science. If you disagree, go meet some of the researchers at NIH. You'll come away impressed.

CREATIVE DESTRUCTION

Creative destruction is one of America's superpowers. If a company is failing, America generally sees no reason to subsidize or otherwise prop it up in order to keep it running.

There are exceptions to this approach—most recently, the American government's response to the Coronavirus pandemic. We also bailed out the auto industry and banks during the Great Financial Crisis.

I believe these are exceptions that prove the rule. In general, we have glorified creative destruction. Even the massive relief package post-COVID won't change this. In such an exogenous, once-in-a-lifetime crisis, the government felt it was in the best interests of the economy and in helping people keep their jobs to support the private sector through a series of subsidies, loans, and grants. In the short term, it was likely the smart choice. However, continuing to support these businesses beyond the immediate crisis would fundamentally go against the American Formula and diminish one of our chief economic advantages, one that has been crucial in fostering a thriving ecosystem of innovation.

In America, the new ruthlessly replaces the old. Blockbuster loses to Netflix, IBM to Microsoft, Blackberry to Apple, etc. The 1990s Meg Ryan and Tom Hanks romcom *You've Got Mail* memorialized the rise of the "big box" bookstore at the expense of the "mom-and-pop" corner store. Just a few years later, Amazon pushed Borders into bankruptcy and left Barnes and Noble on life support. (It's harsh, but I love my Kindle.)

RULE OF LAW

America's legal system can often be frustrating, but it does provide for protection from arbitrary government action and secures commercial rights to property. The ability to protect your property is a key to encouraging investment. In countries where money isn't safe, investors tend to invest in more tangible assets or move that money out of the country into more secure countries. Just see the property market in China and Vancouver for proof.[58] Additionally, our court system and reliance on common law built up over centuries has been a significant strength (and it may be why we love to sue each other so much).

FINANCIAL MARKETS

America has very deep and mature financial markets, which allow entrepreneurs to raise capital and grow large, innovative businesses. That includes bank loans, corporate bonds, private equity, and, of course, venture capital. Our government has little control of the mechanisms that allow investors and the market to allocate capital efficiently.

Venture capital dates back to sixteenth-century England, when investors would pay to outfit trading ships to Asia or the Americas. The captains and crew took a percentage (often

---

58   Chris Rae. "Move Over Toronto and Vancouver—Why Rich Chinese Are Buying Up Montreal Real Estate over Other Canadian Cities." *Style.* February 22, 2020, https://www.scmp.com/magazines/style/news-trends/article/3051767/move-over-toronto-and-vancouver-why-rich-chinese-are.

20 percent) of the profits from the cargo they "carried," and the investors took the rest. "Carried interest" is how today's VCs make most of their money. The model may have been pioneered in Britain, but it was America, beginning with whaling ships in the nineteenth century, that brought the industry into the modern era.

As we saw in chapter 5, it was Arthur Rock, who helped launch Fairchild Semiconductor and Silicon Valley who coined the term. A lot has changed since the days of Arthur Rock. But one thing that *hasn't* is America's continued global leadership in venture capital funding. Part of this deep well of venture capital is due to our innovation, and part of it is due to the strength and maturity of our financial system.

Our liquid and well-regulated public markets are what allow VC investors to exist—and therefore profit from—their investments.

EFFECTIVE AND RELATIVELY LIGHT REGULATION

While this is a deeply contested and political topic, effective regulation has been a key part of the American Formula. We can and must battle over this issue, but it remains true that our regulatory apparatus has allowed for tremendous wealth creation. We don't like how banks have been regulated and bailed out, but it remains true that they are the strongest in the world. It's frustrating that licensing regimes put up barriers to business creation at the state and local level for many professions like beauticians or dog groomers, but we remain an entrepreneurial powerhouse.[59] I would certainly like to see the United States reform many of its regulations, but the fact remains that we remain far more dynamic than do many other parts of the world.

INDIVIDUALISM

The mythology of the lone entrepreneur defying the odds and the powers that be is a powerful one in America. Thomas Edison, Andrew Carnegie, JD Rockefeller, Henry Ford, Steve Jobs, Bill Gates, and Elon Musk were/are not pleasant people particularly when it came to business. While that mythology might be a bit exaggerated, it remains true that American culture rewards success like no other. In many other societies, entrepreneurs are viewed more suspiciously or in the context of their social class. Many societies without a reputation for individualism are entrepreneurial including China's, but I believe the American glorification of individualism gives us an advantage particularly in areas characterized by profound creative destruction and overturning vested interests.

---

59   Jennifer Huddleston. "Could Occupational Licensure Scare Away Entrepreneurship and Innovation?" *The Bridge*. October 30, 2019, https://www.mercatus.org/bridge/commentary/could-occupational-licensure-scare-away-entrepreneurship-and-innovation.

OUR MOST PRESSING QUESTIONS

Considering these strengths, it's little surprise that America has seen such sustained success on the global stage. Every year, the Global Innovation Index uses a series of seven metrics to calculate the most innovative nations in the world. Here is the top ten from 2019:[60]

1. Switzerland
2. Sweden
3. United States of America
4. Netherlands
5. United Kingdom
6. Finland
7. Denmark
8. Singapore
9. Germany
10. Israel[61]

America might not be number one on this list, but the two above us are small, countries that don't rank in the world's top economies. However, as successful as America has been, it hasn't been without its challenges—or without its fault. For us to continue enjoying our status as one of the most innovative nations on the planet, there are four crucial questions we must answer:

---

60 Global Innovation Index. The Global Innovation Index (GII) 2019: Creating Healthy Lives—The Future of Medical Innovation. 2019, https://www.globalinnovationindex.org/home.

61 Just off this list, sitting at thirteen and fourteen respectively, are Hong Kong and China.

- What do we prioritize when several different areas require investment?
- Who will benefit from these investments?
- How do we mitigate the cybersecurity and privacy risks?
- How do we bridge the digital divide?

For the rest of the chapter, we will examine each of these crucial questions in detail.

HOW DOES AMERICA BALANCE ITS MANY INVESTMENT NEEDS?

Both in the 2020s and in the decades that follow, the United States will face several capital-intensive challenges. These include:

- Mitigating the effects of climate change.
- Transitioning to renewable energy sources.
- Reinvesting in America's physical infrastructure (i.e., highways, bridges, and railways).
- Accounting for an increase in healthcare and retirement costs for the Baby Boomer generation.
- Investing in the Fourth IR:
  - AI, VR/AR, IoT, cybersecurity, etc.
  - Biogenetics
  - Digital infrastructure (e.g., aka 5G).

It isn't the purpose of this book to rank these challenges or even to estimate their costs. However, with even this brief list, it is clear that the government's resources will be stretched.

Throughout this book, I have advocated for an urgent and immediate investment from the government in order to build out the nation's 5G infrastructure. But in the midst of these and other challenges, does 5G have a reasonable chance of being funded?

It should. 5G is a core multiuse infrastructure on which many of the solutions to the previously mentioned problems above will be built. Instead of fixing one narrow problem, 5G would create a platform open to a billion applications—many of which would almost certainly be beneficial to our other key challenges and crucial investment needs. 5G will have a tremendous ROI, and it will be crucial to mitigating many of the challenges that lie ahead for both the United States and the world.

IS BIG TECH TAKING OVER?

If we accept that building out our 5G capabilities should be a key strategic priority, what companies will benefit?

The easy answer would be to look to the companies that built out our 4G networks—chiefly AT&T, Verizon, and T-Mobile. Notice that none of those companies appear on the list of most valuable today (see chapter 1). AT&T may have made the list in 2007 when Apple launched the iPhone on its network, but it has slipped into the second decile given the growth of the FAANGs, FANMAG, or whatever silly acronym currently being used to refer to companies like Facebook, Apple, Netflix, Microsoft, Amazon, and Google. The 2010s were dominated by cloud- and 4G-enabled companies like the aforementioned American FAANG companies, as well as by Chinese companies like Alibaba, TenCent, Baidu. etc. As of 2020, these companies are all flush with cash, still growing, and, in the wake of COVID-19, facing less competition than ever.

With fortress balance sheets and thriving (in many cases, dominant) businesses, these companies will gobble up all the talent and smaller companies starved for capital. The FAANGs will benefit enormously from 5G and we will all certainly benefit from their participation. However, if they continue on their path to becoming a kind of corporate oligopoly, their presence will be almost like a tax on the innovation economy and prevent the creative destruction that is our superpower.

This concern is already a major talking point in American politics. Many worry that these companies are either heading toward becoming oligarchies or they're already there. For instance, the Amazon platform allows third parties to sell goods alongside Amazon's own products in their marketplace. This may seem like a healthy free-market principle, but then Amazon uses the data on what those companies are selling to figure out how to sell their *own* versions of those products to the same customers. Similarly, Apple taxes businesses by taking 30 percent off every transaction in its App Store.

I don't blame the companies themselves. They've provided tremendous value, and we all use their products and services because they are good. However, the FAANG companies' continued dominance—and the fact that they are all tech-driven, means that the arrival of 5G will only strengthen their grip on the economy and potentially choke off innovation. We must explore new antitrust policies that balance the benefits these companies have brought to American consumers with the potential stifling of innovation. I also believe that it is imperative that our investments focus on producing open systems and encouraging startup activity across the Nation.

CYBERSECURITY AND PRIVACY

5G is all about data. It allows us to gather more of it and process it more quickly. The mathematician who established Tesco's Clubcard loyalty program once said that "data is the new oil." It's the commodity that makes the digital economy run. But in a *Forbes'* article, the futurist, Bernard Marr, pointed out that unlike oil, data isn't finite. It gets more useful the more you use it, it's cheap to move around, and has myriad applications.[62] In fact, data is far more valuable than oil. Just see those market capitalization comparisons in chapter 1 again.

All the properties that make data so valuable also make it dangerous. The more of our lives that depend on online data, the more vulnerable our privacy becomes. I've personally received at least four or five notices over the last few years of sensitive information of mine that has been hacked from the government, healthcare providers, and others. I can only see that information getting more sensitive and the breaches increasing.

Cybersecurity also poses considerable risks. A fully autonomous driving system is potentially vulnerable to outside hackers. In 2015, white hat hackers were able to operate a Jeep remotely through a backdoor that led to the recall of 1.4 million vehicles.[63] I can't imagine the carnage of a terrorist attack that targeted fully autonomous vehicles or one that shut down our power grids for substantial periods of time.

---

62 Bernard Marr. "Here's Why Data Is Not the New Oil." *Forbes.* March 5, 2018, https://www.forbes.com/sites/bernardmarr/2018/03/05/heres-why-data-is-not-the-new-oil/ - 598378f73aa9.

63 Andy Greenberg. "The Jeep Hackers Are Back to Prove Car Hacking Can Get Much Worse." *Wired.* August 1, 2016, https://www.wired.com/2016/08/jeep-hackers-return-high-speed-steering-acceleration-hacks/.

THE DIGITAL DIVIDE AND WEALTH INEQUALITY

Another problem arises if the private sector alone is left to build out our 5G networks. They will focus all their efforts where the best business case exists—specifically, densely populated urban and suburban areas.

From a private sector perspective, there isn't much of a business case for building out in rural areas. Today, we prize technological skills and concentrate people into wealthier, denser areas, leaving behind those with less education and those who live in rural or nonurban areas. In fact, as of this writing, there are many parts of rural America that the 4G revolution still hasn't reached. When 5G hits, it will bring speeds as much as a hundred times faster than 4G, but only to the most densely populated areas of the country. This would leave rural areas largely without the benefits of 5G and exacerbate the existing broadband gap. It would effectively cut rural America off from the Fourth IR economy, further accelerating the growing advantage gap between urban and rural America and sparking a mass migration from those regions.

We must also realize that it's not only rural areas that will miss out. Big digital divides will form between wealthier urban areas and poorer urban areas, as well as first-tier cities and second-tier cities.

The other question about who benefits concerns wealth inequality. The profits from these companies flow to shareholders and their employees. Both these groups are the already wealthy and/or the well-educated. That leaves large parts of our society out of the equation. 5G will most likely accelerate those inequalities as it draws more jobs away from rural and less connected areas and provides outsized rewards to the shareholders and employees of the big 5G winners.

In chapter 9, I will briefly explore some ways to encourage skilling up workers for the 5G age. It's possible that reshoring jobs from China will help boost employment in traditionally blue-collar professions like manufacturing. However, manufacturing is changing as well. It is increasingly automated and involves robotics. Those jobs are more likely to be in robot programming and maintenance and require technical skills. Skilling up our workforce at every level is the only option we have.

CHAPTER 8

CONTRASTING THE AMERICAN SYSTEM WITH CHINA'S

NOW THAT WE'VE EXPLORED SOME OF THE KEY strengths and some of the big challenges of the American Formula in the previous chapter, let's contrast it to the Chinese system.

China operates differently than the United States. China does have a private sector that complements their state-affiliated sectors. However, with the Chinese system, there's always the question of the rule of law. The Party comes first in China, and if it has to choose either world power status or the power of the Party domestically, the Party will win out every time. As I've mentioned earlier, state-owned enterprises (SOEs) operate in close alignment with the Party. The Party pressures banks to lend to SOEs to maintain employment. That has fueled a mass of empty apartments, ghost cities, and bridges to nowhere, and this becomes particularly pronounced when the economy hits a rough patch.

Rather than enrich their own populace and allow the private sector to allocate resources, the Party decides in many cases. That's why 20 percent of all Chinese apartments are empty even as their prices rise.[64] Of course, that was all fine when China's economy was growing at 15 percent a year. Even at 9 percent, the Party and its cronies took 7 percent, and the workers got only 2 percent. Everyone was getting more, but at 1–3 percent growth or during a contraction like during COVID-19, splitting the pie isn't as easy.

The Soviet Union also struggled to find this balance in the 1980s, but its system was so sclerotic by that time that it simply collapsed. So, what am I worried about? China is misallocating resources, so we can sit back and relax, right? Wrong. Despite its problems, China has far more people and far more resources than Russia. They have the ability to direct the large-scale investments in infrastructure necessary to take the lead in the race to 5G.

---

64   Bloomberg. "A Fifth of China's Homes Are Empty. That's 50 Million Apartments." November 8, 2018, https://www.bloomberg.com/news/articles/2018-11-08/a-fifth-of-china-s-homes-are-empty-that-s-50-million-apartments.

While they can waste money on overbuilt infrastructure, they can also invest productively in 5G at the same time. Further, China also has venture capitalism and a thriving private sector that creaky SOEs have not crowded out. As proof, just look at their ability to create several large technology powerhouses like Alibaba and TenCent. In 2019, the Chinese-owned social network, TikTok, was the most downloaded app in the Apple App Store.[65] They understand what they need to do to win the race to 5G, and they will do what they have to do to get there.

For these reasons—and for all the other factors we discussed in part 1—it is likely that China *will* beat the United States in building out the large-scale infrastructure that will power the Fourth IR.[66] However, if we can be fast followers, then our strengths can help us overcome China's early lead.

Will that happen? There is reason to be optimistic. Every day, we are seeing more signs that the powers that be are waking up to the reality of our present situation and what we need to do to address it. In the wake of COVID-19, we've seen multiple calls for the United States to rebuild as a 5G economy.[67]

---

65   Katie Jones. "Ranked: The World's Most Downloaded Apps." *Visual Capitalist.* January 25, 2020, https://www.visualcapitalist.com/ranked-most-downloaded-apps/.

66   Stu Woo. "In the Race to Dominate 5G, China Sprints Ahead." *The Wall Street Journal.* September 7, 2019, https://www.wsj.com/articles/in-the-race-to-dominate-5g-china-has-an-edge-11567828888.

67   Jeffrey Mervis. "US Lawmakers Unveil Bold $100 Billion Plan to Remake NSF." *Science.* May 26, 2020, https://www.sciencemag.org/news/2020/05/us-lawmakers-unveil-bold-100-billion-plan-remake-nsf.

While it's easy to imagine a future where China does begin to surpass the United States in certain ways, the American Formula gives us certain advantages that may prevent China from staying ahead for long. Both nations are innovative and determined to develop and implement these new technologies, but each has different offsetting advantages.

For their part, China will get out of the gate more quickly because of their authoritarian system. They don't have to worry about environmental reviews, local aesthetics, and they don't have to pass initiatives through Congress, so they can rally their resources quickly.

America will take longer to get moving. However, the American Formula offers us several key advantages (see chapter 7). We have deeper capital markets, the rule of law, and many of the world's top research institutions. Over time, these assets should compound to our advantage, but democracy is messy, and it will take a while, so we have to be prepared for some shocks.

WILL CHINA HAVE A SPUTNIK MOMENT?

On October 4, 1957, the Sputnik satellite became the first-ever human-built object to be launched into the Earth's orbit. Up until that moment, the United States had enjoyed its status as the undisputed innovation leaders of the twentieth century. We developed the first atomic bomb, the first computer, and the vaccine for polio. With Sputnik, the United States had to face an uncomfortable truth:

The Soviet Union beat us.

The government had known the Russians were developing a satellite, just as we were doing the same with its own program. However, we didn't know the Russians were as close to launching as they turned out to be. When we detected Sputnik in orbit, we were caught unprepared.

This moment would become a crucial turning point in America's own space program. Suddenly, the pressure was on to beat the Russians. Early attempts weren't very successful—our first attempted satellite launch in December 1957 resulted in an explosion on the launchpad—and we struggled to find our footing for the next few years afterward. Eventually, then-President John F. Kennedy made the race to the moon a strategic priority with a massive government investment, and the United States landed on the moon in 1969. No other country has since duplicated the feat (and neither have we since 1972).

Both Sputnik and the moon landing had a huge impact on the public psyche. The former brought the devastating news that we had fallen behind a fierce rival, which meant that our spot on the global pecking order was in question. The latter reversed that feeling, and our success made headline news all around the world.

In a historical context, the twelve-year span from 1957 to 1969 is a relatively quick turnaround to go from being caught almost completely unprepared to catching up and then surpassing the program that was beating us. And it wouldn't have been possible without the full commitment of the government in terms of education, investment, and innovation (and, yes, a few former Nazis).[68] The competition got us to invest, and that investment had huge payoffs—for our national security, for our economic and technological supremacy, and perhaps most importantly, for our psyche.

In terms of our Long Competition with China, our Sputnik moment may soon be coming.

It's not difficult to imagine a day when we suddenly realize that the Chinese are far ahead of us in a core technology area. As of this writing, China is already emerging from the COVID-19 pandemic much faster than America and the West. Will they use that time to race ahead while we lick our wounds? For this reason, and the other challenges we will explore in this chapter, the question is how will we respond if that moment comes?

OUR COMPETITORS ARE CATALYSTS

There are two aspects to understanding the Sputnik moment.

---

68 Alejandro de la Garza. "How Historians Are Reckoning with the Former Nazi Who Launched America's Space Program." *Time*. July 18, 2019, https://time.com/5627637/nasa-nazi-von-braun/.

The first was that sense of being caught unprepared. Sputnik's arrival captured the American psyche. People could see the Soviet Union's satellite in orbit in the night sky. They were living in fear, unsure of what Sputnik was, or what this breakthrough might lead to. Given the Cold War atmosphere at the time, both the American government and the American people saw the Soviet Union's accomplishment as a big, crushing defeat. Did it mean that communist central planning was superior to capitalism? Was George Kennan wrong?

The second aspect of the Sputnik moment is what it spurred—chiefly, the creation of NASA, investment in science and math education and new initiatives to double- or triple-down on investments. The Sputnik moment resulted in a massive mobilization of resources. Once again, America leaned all the way into the American Formula we had used so many times before.

During the race to the moon, the government became the primary customer for the nascent space industry. Oddly enough, NASA soon became a service provider for the emergent commercial satellite industry with the launch of the Telstar Satellite in 1962. At the same time, the National Defense Education Act allocated considerable funds to general science and math education to improve the technical skills of a new generation of Americans.

Through these efforts, once again we see the benefits of a government willing to play the long game. The space program created an industry around the government's efforts, but over the long term, it attracted private investment as well. It was a landmark year in 2020 for the public/private partnership in space. In late May, SpaceX became the first commercial company to ferry NASA astronauts to the International Space Station. That achievement was very public, but many missed a potentially more important development. Also in 2020, NASA awarded contracts to three commercial companies: Blue Origin, Dynetics, and SpaceX, to develop a new lunar lander as America attempted to return to the moon.[69] Contracts such as these may signal the dawn of an entire space industry serving both commercial and government needs. That is the American Formula working its magic once again.

But where the space program may be an example of the government playing the long game in the extreme, its investment in the Defense Education Act yielded much quicker results and reinforced America's advantage in top research universities. Whether or not China has its Sputnik moment, we must learn from and replicate the investments we made in the 1950s, 60s, and 70s. But this time, the challenge will be different. In China, we face a different kind of competitor than Soviet Russia.

69   Eric Berger. "NASA Awards Lunar Lander Contracts to Blue Origin, Dynetics—and
     Starship." *Ars Technica*. April 30, 2020, https://arstechnica.com/science/2020/04/
     nasa-awards-lunar-lander-contracts-to-blue-origin-dynetics-and-starship/.

CHINA ISN'T THE SOVIET UNION

With Sputnik, the Soviets were able to direct investment and their own former Nazis into their space program in order to gain an advantage over the United States. It created a short-term victory, but it masked the fact that the Soviet system was fundamentally broken. It didn't have the other parts that make the American Formula so powerful. It had no financial markets to allocate capital and no incentive system to inspire bottom-up innovation and commercial applications. The Soviets failed to create those communities of government, researchers, and commercial businesses, which meant they couldn't stay ahead long term. Once America invested, we were able to run right past them.

China doesn't have the same weaknesses. In fact, they have several advantages that America doesn't:

1. China does not have to worry about privacy and con-sumer protection. In a data-driven economy, this means not having to worry about privacy concerns with inno-vations like AI. Unfortunately, this can be a tremendous advantage.
2. China can roll out a national 5G network without con-cern for the environment or local preferences.
3. China can direct investment including that of the private sector, allowing them to target large amounts of resources for a specific cause. While the United States can direct resources to advance 5G, we face bigger challenges by

either having to subsidize large telecom companies or by competing with them. Either way, our leaders would face political backlash.

4. China has a massive internal market that encourages internal and external investment due to the sheer size of its population and its relative wealth.

These advantages could combine to result in China's Sputnik moment—and that moment could happen in the not-so-distant future. At that point, the United States will realize that we're being outspent and out-innovated—and at risk of losing our position of technological and economic supremacy. With its top-down system and lack of individual rights, China appeared able to contain the spread of COVID-19 quickly. This early containment may allow China's economy to revive faster. Will that head start allow China to outpace the United States in other key technological areas? And if so, in which area will that Sputnik moment come? AI? 5G? Computing?

Imagine what our Sputnik moment will be like if China beats the United States to the quantum computer. In theory, a quantum computer could decode every encryption key in the world in a matter of seconds—rendering all our secret and protected information visible to our biggest competitor. Granted, this is an extreme example, and we haven't done such a bang-up job protecting our information up until now anyway. However, the point remains: a real Sputnik moment with China is possible, and how we prepare for and respond to that moment will shape the trajectory of the Fourth IR.

CHAPTER 9

WHERE DO WE BEGIN?

THE FUTURE IS ABOUT BEING ALMOST THERE NO matter where you are, with advanced telepresence that makes Zoom today look like black-and-white TV. The future is AI that automatically answers your questions before you even ask them. The future is a remote consultation with a medical specialist in San Francisco while you're in Washington, DC, and she can do everything she can do in the office. The future is trusting your driverless car to navigate safely to your destination and make split-second real-time decisions.

This future is all of these things, but only if we start investing higher percentages of GDP in R&D and infrastructure. As I mentioned earlier, the US R&D funding hit a peak of around 2.3 percent GDP in the mid-1960s. However, over the intervening decades, it has slid down below 1 percent to about 0.6 percent. We should at least strive for the 1 percent we achieved with the American Recovery Act after the Great Financial Crisis of 2008 and 2009. One can only hope that the R&D increase discussed for the post COVID-19 recovery gets us there. Unfortunately, our two parties both have other ideas. The right seems to only want tax cuts, and the left only transfer payments. No one wants to build a road—or at least, the road is always each side's second choice.[70]

Once again, we have the opportunity to be leaders in this future and usher in a new American century, but it may require that Sputnik moment to shock our political system out of its familiar ruts. Let's hope we don't have to wait too long. To create the future, we must act quickly and decisively.

But where do we start? Here in our final chapter, we'll examine a framework for using the American Formula to win the Long Competition.

A FRAMEWORK FOR SUCCESS

When taking the bird's-eye view of the Long Competition, it may seem like an impossible, overwhelming task.

---

70   I'm not making a political argument for or against either party's priorities.

However, we don't have any other choice but to try. If we *don't* engage in the Long Competition, if we *don't* remember our great strengths and once again embrace the American Formula for success, then we fall behind.

Fortunately, there's also a lot we can do to make sure that doesn't happen—and it begins with three investments in education, infrastructure, and innovation or as I call it, EI^2. This broad framework allows us to seize upon the best elements of the American Formula and begin to break down our needs in the Long Competition into smaller, actionable pieces.

EDUCATION

According to the World Economic Forum, "65 percent of children entering primary school today will ultimately end up working in completely new job types that don't yet exist."[71] Technology leadership in the Fourth IR begins with education. Many of our institutions of higher learning have been devastated by COVID-19, and our model leading into the pandemic didn't exactly appear that sustainable. Higher education costs have risen eight times faster than wages in recent decades.[72] While the wage premium for college graduates

71 World Economic Forum. The Future of Jobs. 2020, http://reports.weforum.org/future-of-jobs-2016/chapter-1-the-future-of-jobs-and-skills/#hide/fn-1.

72 Camilo Maldonado. "Price of College Increasing Almost 8 Times Faster Than Wages." *Forbes.* July 24, 2018, https://www.forbes.com/sites/camilomaldonado/2018/07/24/price-of-college-increasing-almost-8-times-faster-than-wages/#42061a4766c1.

still tends to make the cost worthwhile, the gain is increasingly eaten up by the colleges themselves.[73]

The cost of goods imported from China might have contributed to low inflation in recent decades, but the cost of services like education, which we do not import from low-wage countries, has risen. This situation has led to a bifurcated experience where the things we want are cheap but the things we need (housing, education, healthcare) grow ever more expensive. As technology accelerates, leaving workers without the ability to refresh their skills is all the more dangerous. Workers need access to continuing education not just for four years, but for their entire professional lives.

The college model will not go away. It still makes sense for a portion of our population. Those institutions are also the backbone of our research and development capability in the United States, so I'm not calling for their shutdown or impoverishment. What we need is a system for the rest that allows us to skill up and retrain workers over the course of their careers. Our technical schools and community colleges need a massive overhaul. We also need to explore the German model, where the government and companies share the costs of employees during an internship or apprenticeship period

73 Michelle Singletary. "Is College Still Worth It? Read This Study." *The Washington Post.* January 11, 2020, https://www.washingtonpost.com/business/personal-finance/is-college-still-worth-it-read-this-study/2020/01/10/b9894514-3330-11ea-91fd-82d4e04a3fac_story.html.

that includes work and classroom training.[74] We also might need to cycle through several periods of apprenticeship over our careers. This goal conflicts with the very premise of our current system of education as an early life activity.

5G itself will enable some of these shifts and ensure they are affordable. Although online education doesn't feel great today, the advancements in telepresence and virtual reality will make educating learners remotely a far more "in-person-like" experience. We can educate workers in shorter and more effective bursts. We can also tailor education more individually at scale. The techniques that work for an eighteen-year-old valedictorian will not be the same ones that are most effective for a fifty-five-year-old mechanic learning to maintain a new fleet of autonomous vehicles. Additionally, 5G technology will allow instructors to leverage their time and salaries across more learners and get better results at a far lower cost. These same benefits should flow down to our secondary schools as well. A growth in educational productivity is sorely needed, and without it, we may not win the Long Competition.

Congress must find common ground to make the investments to enable better and more ongoing technical education and to revolutionize instruction through 5G technology. If

---

74   Tamar Jacoby. "Why Germany Is So Much Better at Training Its Workers." *The Atlantic*. October 16, 2014, https://www.theatlantic.com/business/archive/2014/10/why-germany-is-so-much-better-at-training-its-workers/381550/.

we can do this, the return on investment will be enormous. We'll dwarf the incredible gains from last century's GI Bill that educated the Great Generation returning from World War II, which from the end of the war through 1956, helped educate sixteen million Americans and was one of the drivers of our post WWII prosperity.

In the words of the historian, Ed Humes, "The GI Bill provided the education for fourteen Nobel Prize winners, three Supreme Court justices, three presidents, a dozen senators, two dozen Pulitzer Prize winners."[75] In the wake of COVID and in the face of a looming competition with China, we need a similar approach. Without these investments, we risk creating a shortage of qualified labor and a disenfranchised underclass made up of the noncompetitive underemployed.[76]

INFRASTRUCTURE

The United States must also undertake an infrastructure program spanning from bridges to rail to broadband. Our infrastructure gap is huge. In 2017, the American Society of Civil Engineers estimated that America would need $2 trillion in additional infrastructure investment over the

---

75 Marketplace. "How the GI Bill Changed the Economy." October 6, 2009, https://www.marketplace. org/2009/10/06/how-gi-bill-changed-economy/.

76 I am also not a fan of universal basic income. Work is important to humans, and I think UBI would lead to widespread despair and societal breakdown.

next ten years.[77] Okay, maybe that's a bit like asking your barber if you need a haircut, but the number is big nevertheless. The estimated investment in digital 5G infrastructure adds up to another estimated $2.7 trillion.[78] Of course, that figure includes much-needed private sector investment as well. These numbers are rough and likely a bit inflated, but it gives us a jumping-off point. The dollars needed to remain competitive are huge.

To make this investment politically and financially palatable, we need a variety of mechanisms and funding sources. First, we can push some funding down existing federal grant and loan program channels through the federal bureaucracy. This approach makes sense for traditional infrastructure areas where we have programs already set up, such as the Federal Highway Administration. It also makes sense to use grants and loans through federal agencies when the government has a particular interest that the commercial sector does not want to support. Loans and grants to provide rural broadband access might fall into this category.

Additionally, we must also provide substantial funding to states and localities to work in partnership with industry to build out smart cities/communities infrastructure at the

---

77 The American Society of Civil Engineers. "2017 Infrastructure Report Card," 2017, https://www.infrastructurereportcard.org/solutions/investment/.

78 Macy Bayern. "Why a 5G Rollout Requires $2.7T Investment by 2020." *TechRepublic*. February 25, 2019, https://www.techrepublic.com/article/why-a-5g-rollout-requires-2-7t-investment-by-2020/.

local level. These efforts could include sensor and camera networks, AI capabilities, and infrastructure for autonomous vehicles. IoT infrastructure at the local level will allow thousands of entrepreneurs to build new businesses on top of it. It's essential that this infrastructure remain as open as possible to allow these innovators to thrive. If we can keep the systems open and out of the control of the existing tech giants, a large state and local IoT investment could create a new entrepreneurial wave across the country.

Finally, we also need to explore a federal infrastructure bank that could loan funds at low rates and on generous terms to states, localities, and commercial enterprises to pursue national priorities. An infrastructure bank would attract additional capital, financial discipline, and a focus on return on investment from the private sector. It also allows for investment in broad areas outside of the priorities of a specific federal department.

Structuring a bank or banks and their governance will be contentious, but they could be powerful mechanisms to use federal seed capital to tap national wealth and expertise. If the government provides some of the seed capital or loan guarantees, it helps remove risk from a project. That reduces the return required by private investors and makes many more projects feasible that might otherwise be considered too much risk for too little return.

INNOVATION

Like the rest of us, the government does not know what the next generation of 5G applications might look like. However, we do know that several key technologies will likely be important, such as AI, biogenetics, robotics, autonomy, additive manufacturing, IoT, VR/AR, etc. The government can help fund a new generation of innovators in our universities and research institutions and use modern acquisition methods, such as prize challenges and Other Transaction Authorities (OTA), to draw new innovators and companies into emerging technology areas.

As we saw in part 2, the government has a history of supporting innovation—including Hamilton's bounties, Congress's funding of the telegraph, and the internet today. In recent decades, a prize challenge from DARPA launched the autonomous vehicle revolution and the Human Genome Project launched a new era in biogenetics.

Investment in innovation is not a one-size-fits-all strategy. We need to "let a thousand flowers bloom."[79] Some funding should go to government agencies to invest in their mission areas from defense to energy to health and human services. These agencies are better at applying research to specific challenges associated with their missions. Some should go to government research organizations like the National Science Foundation, NIST, NIH, and DoE that are pushing

79   To paraphrase Chairman Mao.

the boundaries of knowledge in more fundamental research areas. These agencies in turn seed many of our research universities and institutions.

The government has also been experimenting with partnerships with venture capitalists and accelerators/incubators who can help determine which ideas and companies have a shot at creating sustainable businesses. While these efforts are relatively new and need to be studied further, the general idea of the government either getting out of the way or smoothing the path to commercialization from research is a key priority. Again, the government should not pick the winners and losers, but they can encourage those who should (investors) to come to the table.

Finally, the government should ensure a segment of funding goes to smaller and startup companies. Policy makers can build off the existing Small Business Innovation Research (SBIR) program and create new ways for startups and smaller companies to work directly with the government without having to contract through an existing prime contractor or take on the full costs of complying with federal contracting mandates. SBIR programs today are often an afterthought, but if we can modernize them and scale up the funding, it could become a driver of more small and innovative companies.

THE FUTURE IS OURS

If we can do that, the reward won't just be winning the Long Competition. It will be winning a better way of life.

Think of how far we've come in just two and a half centuries since the founding of the United States. We started as a weak collection of colonies dependent on agricultural commodities. Two hundred and fifty years later, we are a continent-spanning nation able to project power anywhere on the globe. With each previous industrial revolution, we have seen innovations that have improved the quality of human life. They have lifted more people out of poverty, extended lifespans, and produced the greatest accumulation of wealth in human history. However, our work is not done.

Now, think of what we can accomplish in this latest revolution. The technologies and applications of a 5G-enabled society will have far-reaching benefits for society. A move to electric vehicles, for instance, will likely correspond with a move to alternative energy sources, which will help mitigate the effects of climate change. Biogenetics offers solutions to once chronic or fatal diseases. Other innovations will help us tackle crucial health issues, increase wealth, and create new opportunities for people to learn, express themselves, and build businesses.

That's the future we're fighting for—so let's get started creating it.

CONCLUSION

TOWARD THE END OF THE SECOND INDUSTRIAL REVO-
lution, America began to lay the groundwork for its highway
system. It all began with a cross-country military convoy in
1919—where, along for the ride, was future President and
World War II hero, Dwight D. Eisenhower.

The purpose of the convoy was to bring attention to how bad
America's roads were. Armed with a Militor, a vehicle capable
of dragging cars out of ditches, they set out on their way. Even
with the extra lift, it still took them two full days to travel the
forty-five miles from Lafayette Park in Washington, DC, to
nearby Frederick, Maryland (and they couldn't blame it on
the traffic back then). From there, they set out west, eventually getting all the way out to Wyoming—where the roads
were so poor (and often nonexistent) that even the Militor
got stuck. By the end of the over sixty-day journey, they had

made it all the way to the West Coast. The effort made it clear that America desperately needed to modernize its roads for the era of the automobile, but sadly, no one was listening.

The convoy may have failed at its mission to generate awareness for America's inadequate roads, but it left an impression on the young Dwight Eisenhower. When he became president decades later, Eisenhower committed the country to its greatest investment in the highway system up to that point.

It's commonly believed that the highway system was largely built for military use, but that isn't the case. From its inception, the highway system was intended for commercial use—though the DoD did intervene to make certain secondary military provisions. In truth, because the states matched federal funds—and because states don't spend on defense—the highway system necessarily had to serve commercial purposes by design.

Once again, with this massive government investment, the private sector was able to expand in new ways. With the entire country now interconnected, interstate shipping and commerce became easier than ever, and America's economy grew to the largest in the history of the world.[80] [81]

---

80 Lori Van Pelt. "Eisenhower's 1919 Road Trip and the Interstate Highway System." *WyoHistory.org.* January 4, 2018, https://www.wyohistory.org/encyclopedia/eisenhowers-1919-road-trip-and-interstate-highway-system.

81 Christopher Klein. "The Epic Road Trip That Inspired the Interstate Highway System." *History.* October 28, 2018, https://www.history.com/news/the-epic-road-trip-that-inspired-the-interstate-highway-system.

Today, as we consider our place in the Long Competition, we face a crucial choice: Do we invest or continue with business as usual? Do we build the highway system of the future otherwise known as 5G at scale, and as quickly as possible, or do we cede leadership to China?

In a world where China gained long-term economic advantage, the United States—and much of the world—would find itself poorer and less able to shape the world in its image. We would lose leverage on the global stage and find it more difficult to compete. We would even be at risk of losing much of the considerable wealth we have amassed.

These are all real threats to America's way of life, and we should be aware of them. However, we can't let this awareness devolve into fear. We can't let it be the primary motivator for what we set out to do next. Instead of focusing on what we stand to lose in the Long Competition, we should ask ourselves what we *gain*.

In the introduction to this book, we started with the story of the early days of the autonomous vehicle. The very term "driverless car" elicits visions of a sleek and efficient sci-fi world.

That may one day be the reality. But we need to remember that the driverless car came from humble beginnings. It came from a dry California desert. It came from grit, determination, and endless amounts of trial and error. It came from failure. But most importantly, it came from investing in the future.

When I think of the American Formula, I consider this commitment to the future to be its greatest by-product. At times we take this for granted, but we need to remember that until the First Industrial Revolution, the average lifespan for humans was thirty-five.[82]

Technological progress since then has not only more than doubled the average lifespan, but it has also lifted billions of people out of poverty, including hundreds of millions of Chinese citizens. A pauper today lives better than a king three hundred years ago. Technology is the driver of progress both in wealth and quality of life. Many of the challenges we face as a society, from pandemics to climate change, require *more* technology, not less. Technological progress is certainly key to winning the Long Competition, but it's also crucial to the prosperity and happiness of future generations.

ABOUT CORNER ALLIANCE

Corner Alliance's core purpose is "To help government do amazing things." We believe that partnership between the

---

82 Paul Hodges. "Rising Life Expectancy Enabled Industrial Revolution to Occur." Independent Commodity Intelligence Services. February 27, 2015, https://www.icis.com/chemicals-and-the-economy/2015/02/rising-life-expectancy-enabled-industrial-revolution-to-occur/.

government and the private sector is the key factor in bolstering our national competitiveness. We've been doing that since 2007. Since then, Corner Alliance has worked across agencies such as the Departments of Homeland Security, Agriculture, Commerce, Defense, and the National Institutes of Health to establish and implement grants, loans, prize challenges, and other programs to spur innovation. We've helped our clients and their customers identify essential technology areas to invest in and show the impact and return on those investments.

We exist to help government agencies employ the American Formula to spur national competitiveness and create a better future.

If you're interested in further information about the ideas in the book or in working with Corner Alliance, please contact us at apentz@corneralliance.com or connect with me on LinkedIn (@apentz).

Together, we will win the Long Competition.

ACKNOWLEDGMENTS

I'D LIKE TO THANK ALL MY FELLOW DUCKS AT CORNER Alliance who have helped create an amazing company, as well as all of our clients who make our work possible.

I also owe a debt of gratitude to my wife, Kerri, who had to listen to all my half-baked ideas and halting lectures on electrifying topics, such as foreign currency flows. Whatever debt she owes the universe has been paid in full. I love my kids, Sophie, Sam, and Hazel, but they got to watch YouTube and play video games once I started talking.

I'd also like to thank Chas Hoppe, who helped me with organizing and writing this book. He had to listen to the same ideas, but at least he got paid for it.

Finally, I'd like to thank a bunch of people I've never met, but who did so much to educate me: John Mearsheimer for an education in the realist school, Michael Pettis on the aforementioned currency flows and China's economy, Bill Bishop with his newsletter Sinocism, Ben Thompson with his newsletter Stratechery, and Peter Zeihan for his book *Disunited Nations.*

ABOUT THE AUTHOR

ALAN HAS WORKED WITH GOVERNMENT LEADERS IN the R&D and innovation communities across DHS, Commerce, NIH, state and local government, and the nonprofit sector, among others. He has worked in the consulting industry for over ten years with Corner Alliance, SRA, Touchstone Consulting, and Witt O'Brien's. Before consulting, Alan served as a speechwriter and press secretary for former US Senator Max Baucus and as a legislative assistant for former US Representative Paul Kanjorski. He holds an MBA from the University of Texas at Austin.

Alan is a lifelong fan of the Philadelphia Eagles (yes, they booed Santa Claus), and his hero is Benjamin Franklin, a famous polymath, patriot, diplomat, scientist, politician, and still the founding father you most want to have a beer with.